SpringerBriefs in Political Science

SpringerBriefs present concise summaries of cutting-edge research and practical applications across a wide spectrum of fields. Featuring compact volumes of 50 to 125 pages, the series covers a range of content from professional to academic. Typical topics might include:

- A timely report of state-of-the art analytical techniques
- A bridge between new research results, as published in journal articles, and a contextual literature review
- A snapshot of a hot or emerging topic
- An in-depth case study or clinical example
- A presentation of core concepts that students must understand in order to make independent contributions

SpringerBriefs in Political Science showcase emerging theory, empirical research, and practical application in political science, policy studies, political economy, public administration, political philosophy, international relations, and related fields, from a global author community.

SpringerBriefs are characterized by fast, global electronic dissemination, standard publishing contracts, standardized manuscript preparation and formatting guidelines, and expedited production schedules.

Melissa Schnyder

Global Norms in Local Contexts

Examining Cases of Environmental
Governance in France

 Springer

Melissa Schnyder
School of Security and Global Studies
American Public University System
Charles Town, WV, USA

ISSN 2191-5466   ISSN 2191-5474   (electronic)
SpringerBriefs in Political Science
ISBN 978-3-031-41107-6   ISBN 978-3-031-41108-3   (eBook)
https://doi.org/10.1007/978-3-031-41108-3

This Springer imprint is published by the registered company Springer Nature Switzerland AG
The registered company address is: Gewerbestrasse 11, 6330 Cham, Switzerland

Paper in this product is recyclable.

To Sylvia and Gigi

Preface

The inspiration for this project was largely drawn from the Cities for CEDAW campaign. The Convention on the Elimination of All Forms of Discrimination Against Women (CEDAW), which sets forth principles of fundamental human rights for all the world's women and girls, is a landmark international agreement that was adopted by the United Nations (UN) General Assembly in 1979. In 1998, norm entrepreneurs and activists in San Francisco initiated the idea of drawing upon CEDAW to enhance the status of women locally. Activists were successful in getting a binding ordinance passed that integrated the principles of CEDAW into local governance. This was a pivotal initiative, as it set the ball rolling for other cities in the United States to follow suit and pass similar CEDAW legislation or nonbinding resolutions at the local level, including Los Angeles, Berkeley, and Daly City, California; Louisville, Kentucky; Kansas City and University City, Missouri; and Cincinnati, Ohio. As of today, many cities have undertaken efforts to integrate the principles of CEDAW at the municipal level. In recent years, we have observed similar developments in the area of climate change, as some cities have adapted principles and recommendations from the UN Climate Change Conferences into their own local climate plans to respond to the global climate crisis, such as Climate Ready Oak Park, which was formally adopted by the Village Board in Oak Park, Illinois in 2022.

Although we can observe examples like these of local stakeholders adapting global-level norms and principles to local contexts, we still know relatively little about how global norms are translated into local settings in different country contexts. Thus, this book is an effort to better understand the dynamics surrounding how this process of norm adaptation unfolds in an advanced democracy. Through an exploration of three cases in France (representing a most-likely case design that sets it apart from much of the localization research), the book develops a theoretical framework that combines concepts from the localization literature with processes from the literature on norm-based institutional change, which can help make sense of how local stakeholders translate global norms into local practices. The book's focus is on twilight norms – a specific set of global-level environmental norms that have been codified in international treaties but have not yet been systematically examined at the sub-national level. Given the urgency of the global climate crisis and the lackluster response of many national governments, examining how environmental conservation and sustainability norms are adapted and translated by local actors into local contexts represents an important and timely area of research. In short, it brings the Cities for CEDAW motto "Bring global local" into sharper focus.

Since it is a theoretically informed study, this book will be of interest to students and scholars interested in international norms, normative change, norm diffusion processes, civil society activism and advocacy, and/or current debates surrounding environmental sustainability and conservation. However, it is also for practitioners and activists, as well as members of the public interested in localizing global norms into their communities.

There are many individuals who made it possible to create this book. It is first necessary to thank the local key informants who gave their time to make this research possible. It is also necessary to thank American Public University for providing two research grants to support the fieldwork, which was initially disrupted due to the COVID-19 pandemic. I also extend a sincere thank you to the editors and designers at Springer Nature who have been extremely helpful, and for whose assistance I am very grateful. Finally, I thank my family for their continual support of my research projects, and for their willingness to embark on the adventures that come along with fieldwork across multiple cities and countries.

Charles Town, WV, USA   Melissa Schnyder

Contents

1 Introduction: Examining Global Environmental Norms in Local Settings 1
  1.1 Research Questions and Aims 2
  1.2 Global-Level Environmental "Twilight" Norms 3
  1.3 Framing the Problem 7
  1.4 Local Actors.. 9
  1.5 Multi-stakeholder Governance and Environmental Sustainability 10
  1.6 France as a Backdrop..................................... 11
  1.7 Research Design: Examining Three Local-Level Cases 12
      1.7.1 The Cerbère-Banyuls Marine Nature Reserve............ 12
      1.7.2 The Thau Fisheries Local Action Group (FLAG) 13
      1.7.3 The Biovallée Biodistrict 13
  1.8 Materials and Methodology................................ 15
  1.9 Contributions of the Research 16
  1.10 Plan of the Book... 16
  References... 17

2 How Global Norms Travel: A Theoretical Framework 21
  2.1 Norm Diffusion 21
  2.2 Localization Dynamics 24
  2.3 Issue Framing... 27
  2.4 Actor Constellations...................................... 29
  2.5 Norms and Impact Translation............................. 30
  2.6 Summary and Conclusion 30
  References... 31

3 The Cerbère-Banyuls Marine Nature Reserve................... 35
  3.1 Background and Actor Constellations 36
  3.2 Environmental Norms 38

    3.3   Framing 41
    3.4   Conclusion 43
    References 45

4   The Thau Fisheries Local Action Group 47
    4.1   Thau FLAG Actor Constellations 48
    4.2   Global Environmental Norms 50
         4.2.1   Sustainable Development 50
         4.2.2   Norm Adaptation: Normative Reframing,
                  Grafting, and Norm Linkages 52
    4.3   A Matter of Preserving and Promoting Cultural Heritage 55
    4.4   Conclusion 57
    References 58

5   The Biovallée Biodistrict 61
    5.1   Actor Constellations in the Biovallée Biodistrict 63
    5.2   Global Twilight Norms 65
         5.2.1   Sustainable Development and Intergenerational
                  Equity 65
    5.3   Norm Adaptation Through Foregrounding
         and Normative Innovation 67
         5.3.1   New Concepts and Transformed Perceptions 69
    5.4   Conclusion 70
    References 71

6   Conclusion: Looking Back and Ahead 73
    6.1   Research Questions, Key Findings, and Discussion 73
    6.2   A Focus on the Sub-National Level 76
    6.3   Relevance for Other Contexts and for Future Research 77
    References 79

About the Author

Melissa Schnyder is a faculty member in the Doctorate of Global Security program within the School of Security and Global Studies at American Public University. Her research focuses on transnational social movements, human security policy issues, and European Union politics, with a focus on the political participation and influence of non-state actors. As a former Fulbright Fellow to the European Union, she was based at the Institute for European Studies in Brussels, Belgium, where she conducted extensive fieldwork and interviews with policymakers in the European Commission, elected officials in the European Parliament, and NGO practitioners. She has been a visiting researcher at universities in Belgium and France, has assisted with research projects at the German Marshall Fund of the United States' Transatlantic Center in Brussels, and has been invited to speak at transnational symposia in Europe and the United States. Her research has been published in a variety of international peer-reviewed journals. Past monographs include *Activism, NGOs, and the State: Multilevel Responses to Immigration Politics in Europe*, and *Advocating for Refugees in the European Union: Norm-based Strategies by Civil Society Organizations* (with Noha Shawki).

Chapter 1
Introduction: Examining Global Environmental Norms in Local Settings

In 1984, the Fédération des Associations pour le Développement de l'Emploi Agricole et Rural (FADEAR) was created in France under the auspices of farmers wishing to develop an agricultural model according to which they could earn a decent living from their work. The new model that they envisioned centered heavily on the value of respect for nature. Drawing upon the direct experience and knowledge of farmers, and in partnership with researchers, a number of working groups were created in the process to further develop and define these ideas though the creation of a Charter for Peasant Agriculture. This Charter, which represents many years of collaborative work between farmers and researchers to settle on an agricultural orientation that would allow farmers to work in a sustainable way, was finalized in 1998 by FADEAR and then adopted during of the Rambouillet Congress of the Peasant Confederation (FADEAR, n.d.-a). The Charter is notable, for one, in that its roots lie in a sub-national, farmer-centered initiative. In addition, it is notable for incorporating 10 fundamental political principles (see Table 1.1) designed not only to guide daily farming practice and analysis of the agricultural world on the part of the farmer, but also to guide decision-making by political elites (FADEAR, 2021).

Although the Charter is not international in scope, the global-level environmental norms of *intergenerational equity*, *common heritage of humankind*, and *sustainable development* feature prominently throughout. For example, as part of the 10 principles, principle 3 highlights respect for nature and the climate, explaining that "To produce, agriculture uses the elements of the natural environment: water, soil, air, biodiversity. These elements, which constitute the basis of the work of peasants, are the good of all. They are not the property of a generation of peasants. The natural elements must be preserved, in order to ensure the sustainability of their use by future generations" (FADEAR, n.d.-b, p. 2). In general, the Charter places great emphasis on environmental stewardship in terms of the long-term viability of nature. As it explains, "In order to enable future generations to meet their own needs, the

© The Author(s), under exclusive license to Springer Nature
Switzerland AG 2023
M. Schnyder, *Global Norms in Local Contexts*, SpringerBriefs in Political
Science, https://doi.org/10.1007/978-3-031-41108-3_1

Table 1.1 The ten principles of the Charter for Peasant agriculture

Number:	Principle:
1	Distribute production volumes in such a way as to allow as many people as possible to access the profession and make a living from it.
2	Apply the concept of food sovereignty at home and elsewhere.
3	Respect nature and the climate.
4	Valorize abundant resources and spare scarce resources.
5	Seek transparency in the acts of purchase, production, processing, and sale of agricultural products.
6	Ensure for everyone the good taste and healthful quality of the products.
7	Aim for maximum autonomy in the operation of farms.
8	Seek partnerships with other stakeholders in the rural world.
9	Maintain and improve the diversity of animal populations and cultivated plant varieties.
10	Always think in the long term and in a global way.

Source: Adapted from "Translations of the French Charter for Peasant Agriculture," retrieved from https://ec.europa.eu/eip/agriculture/sites/default/files/fadear_1998_charte_paysanne_eng.pdf

preservation of natural resources, heritage and the environment is a priority that agricultural systems must take into account" (FADEAR, n.d.-c, p. 8). How is it that global-level environmental norms feature so prominently in an initiative that, at its roots, is very much centered on local actors and local territory?

1.1   Research Questions and Aims

This research is concerned with how global-level environmental norms travel – not horizontally across countries or between international institutions, but rather vertically across levels of governance, from the global to the local. It asks how global environmental norms are (re)interpreted by local-level actors and translated to a particular local context. More specifically, this study assesses the significance of so-called global environmental twilight norms (described below) by examining the extent to which, and how, they have been adapted to three local contexts in France, each involving multi-stakeholder environmental governance: (1) the Cerbère-Banyuls Marine Nature Reserve, (2) the Thau Fisheries Local Action Group (FLAG), and (3) the Biovallée biodistrict. In each of these cases, the research assesses how twilight norms are used to frame, promote, and generally develop a local discourse that centers on environmental conservation and sustainability. The specific research questions ask: How do local actors incorporate and promote global environmental norms to govern natural resources in a sustainable manner? Which of the global twilight norms are applied? How are these global norms adapted to local settings? How are global twilight norms used in issue framing, to frame environmental conservation and sustainability in the local context? In general, how do these global norms contribute to the development of a discursive field involving environmental sustainability at the local level? The study is thus concerned with the extent to which global-level environmental norms have diffused to the local level, and how global-level environmental norms are adapted to local settings.

1.2   Global-Level Environmental "Twilight" Norms

The types of global environmental norms of focus in this research are what are sometimes referred to as "twilight norms" in international environmental law (IEL) (Beyerlin, 2008). Table 1.2 lists and defines each twilight norm examined in this study.

Twilight norms represent an interesting focal point of analysis, as they have been said to occupy a place toward the bottom of the normative hierarchy of contemporary international environmental law (Beyerlin, 2008; Kotzé & Muzangaza, 2018), suggesting that these norms are of lesser normative quality than others and may

Table 1.2   Types and definitions of environmental twilight norms in international environmental law

Norm:	Definition:
Precaution	Action should be taken to protect the environment when scientific uncertainty exists (see McIntyre & Mosedale, 1997; Kravchenko et al., 2012; Beyerlin, 2008).
Polluter pays	Those who pollute the environment are responsible for bearing the costs of the damage and must repair that damage (see Kravchenko et al., 2012; Beyerlin, 2008; Nanodkar, 2018).
Common but differentiated responsibilities	The responsibility for addressing global environmental destruction lies with all nation-states in common, yet not all nation-states bear equal responsibility, with developed countries obligated to bear a greater share of the burden than developing countries (see Stone, 2004; Beyerlin, 2008; Kravchenko et al., 2012).
Equitable utilization of shared natural resources	Nation-states must cooperate in the management of shared natural resources when those natural resources form a common boundary between nation-states, or are located within more than one country. Each of the affected nation-states has an equitable share in the use of the resource (see Beyerlin, 2008; Bilder, 1980; Tanzi, 2020).
Intergenerational equity	The environment and natural resources should be used in a way that ensures their preservation for the benefit of future generations (see Beyerlin, 2008; Kravchenko et al., 2012; Rosencranz, 2003).
Common heritage of humankind	Certain natural resources and parts of the natural environment are the property of humankind as a whole (i.e., the global commons) and therefore cannot be unilaterally exploited by a specific nation-state or corporation (see Qureshi, 2019; Beyerlin, 2008; Baslar, 1998).
Sustainable development	In the context of economic growth, the environment must be preserved for future generations while meeting the needs of the present (see Kravchenko et al., 2012; Barral, 2012).

Sources: Compiled from Barral (2012), Baslar (1998), Beyerlin (2008), Bilder (1980), Kravchenko et al. (2012), McIntyre and Mosedale (1997), Nanodkar (2018), Qureshi (2019), Rosencranz (2003), Stone (2004), and Tanzi (2020)

occupy a "gray area" between hard and soft law (Beyerlin, 2008). As a result, there is considerable confusion concerning the role that these twilight norms play in environmental governance and the effect that they have (Beyerlin, 2008). This confusion presents an opportunity to study the impact of these norms in different contexts. Although the extent of their impact remains contested, what is clear is that each of these norms has a long history of embodiment in different treaties and cross-cutting issue areas in international law. Although some have been and continue to be controversial at the international level (e.g., common but differentiated responsibilities), all are recognized by the international community and applied in legally-binding treaties and other soft law instruments.

For example, at the international level, the *precaution* norm was first explicitly discussed and applied in 1987 in the Second International Conference on the Protection of the North Sea, in the context of taking action to protect the North Sea even before clear scientific evidence could be established to indicate damage or destruction (Kravchenko et al., 2012). Several years later, with the 1990 Bergen Ministerial Declaration on Sustainable Development, national ministers agreed to abide by the rules of the precautionary principle (Kravchenko et al., 2012). Part of this agreement entailed that when damage to the environment is imminent, nation-states cannot use a lack of scientific evidence as a justification to delay taking actions to prevent environmental destruction (United Nations, 1990). Examples of other international treaties that apply the norm of precaution include the 1990 Montreal Protocol on Substances that Deplete the Ozone Layer, the Protocol on Water and Health of the 1992 Convention on the Protection and Use of Transboundary Watercourses and International Lakes, the 1992 Rio Declaration, the 1994 United Nations Framework Convention on Climate Change, and the 1992 Convention on Biological Diversity (Kravchenko et al., 2012). These and other examples underscore this norm's wide recognition and acceptance by the international community and reflect its incorporation into different international legal instruments.

The *polluter pays* norm, which assigns responsibility for pollution and other environmental damage to those responsible for causing it, has a similarly long history. At the international level, the 1972 Council Recommendation on Guiding Principles Concerning the International Economic Aspects of Environmental Policies of the Organisation for Economic Cooperation and Development (OECD) represents the first explicit mention of the polluter pays norm. According to the Council Recommendation,

> the polluter should bear the expenses of carrying out the … measures decided by public authorities to ensure that the environment is in an acceptable state. In other words, the cost of these measures should be reflected in the cost of goods and services which cause pollution in production and/or consumption. Such measures should not be accompanied by subsidies that would create significant distortions in international trade and investment. (OECD, 1972, para. A.4.)

In the 1990 International Convention on Oil Pollution, Preparedness, Response and Cooperation, the polluter pays norm was acknowledged as a general principle of international environmental law (Kravchenko et al., 2012). There are many examples of international treaties and agreements that embody the polluter pays norm, demonstrating its acceptance by the international community as a basis on which to protect the environment. A few notable examples include the 1985 ASEAN Agreement on Conservation on Nature and Natural Resources, the 1986 Single European Act regarding the environment, the 1992 Convention on the Protection of the Marine Environment of the Baltic Sea, and the 1992 Convention on the Transboundary Effects of Industrial Accidents (Kravchenko et al., 2012).

The norm of *common but differentiated responsibilities* rests on the dual premise that all states share a common obligation to ensure protection of the environment, but their relative contributions to addressing environmental problems will vary depending upon a particular state's specific resources, situation, and overall capabilities (Kravchenko et al., 2012). It is a norm that is widely reflected in international treaties and other instruments, particularly within the global climate change regime (Wang & Gao, 2018; Yamin & Depledge, 2004). The 1972 Stockholm Declaration on the Human Environment reflects the idea of common but differentiated responsibilities in acknowledging that some standards that are appropriate for some countries may impose undue costs on others (United Nations, 1973, Principle 23). This norm is also reflected in the Rio Declaration, the Montreal Protocol on Substances that Deplete the Ozone Layer, the 1992 UN Framework Convention on Climate Change, and the subsequent Copenhagen Accord (Kravchenko et al., 2012).

The *equitable utilization of shared natural resources* is another norm that has become evident in international environmental law. This norm denotes a cooperative management approach between states when dealing with natural resources that are shared in common between them, and is reflected in hundreds of international agreements that deal with the management of lakes, rivers, and drainage basins, including the 1909 United States-Canadian Boundary Waters Treaty (Bilder, 1980). Moreover, the equitable utilization norm has also been applied in cases involving the cooperative management of fisheries, as well as to problems of international pollution (Bilder, 1980). Legally binding international agreements such as the 1972 London Convention on the Prevention of Marine Pollution by Dumping of Wastes and Other Matter, the 1974 Convention on the Protection of the Marine Environment of the Baltic Sea Area, and the 1976 Convention on the Protection of the Mediterranean Against Pollution help illustrate this norm's broad acceptance and basis as a framework of cooperation in managing shared resources.

Also widely accepted in international law is the norm of *intergenerational equity*, which recognizes the interests that future generations have in the natural environment and their status as legitimate beneficiaries of environmental protection (Weiss, 1996). It stipulates that humankind cannot leave the environment in a worse state for future generations than that in which they have experienced it themselves, and thus aims to prevent irreversible environmental damage (Weiss, 1996). A very early appearance of the intergenerational equity norm is found in 1946 in the International Convention on the Regulation of Whaling, whose preamble upholds the "interest of the nations of the world in safeguarding for future generations the great natural resources represented by the whale stocks" (International Convention for the Regulation of Whaling, 1946, preamble). Since that time, the norm has appeared in many binding international instruments as well as soft law, including the 1972

Stockholm Declaration on the Human Environment, the 1975 Convention on International Trade in Endangered Species of Wild Fauna and Flora (CITES), the United Nations Economic Commission for Europe Water Convention, the 1992 Convention on Biological Diversity, and the 1992 United Nations Framework Convention on Climate Change (Kravchenko et al., 2012).

Common heritage of humankind is a norm that has its foundations in ancient Roman law (Qureshi, 2019). It stipulates that because certain natural resources, such as oceans and seas, are commonly owned by humanity as a whole, no individual, organization, or state may exclusively exploit or own such resources (Qureshi, 2019). The principle of common heritage dates back to a 1967 speech given by the former ambassador of Malta to the United Nations, in which he emphasized the need to take stringent measures to declare the deep seabed and its resources as belonging to the common heritage of the whole of humankind (Akintoba, 1996). This principle was to help allay fears among developing states that developed states would use their technological infrastructure to extract all resources, leaving nothing for poor states (Qureshi, 2019). This norm is embodied in international law including the Geneva Law of the Sea, and it "now also considers Antarctica, other planets, the moon, and human genomes, as well as genetic resources attained from plants and microorganisms, as the common heritage of the whole of mankind" (Qureshi, 2019, p. 86; see also Taylor, 2019).

Lastly, the *sustainable development* norm is the final twilight norm of focus here. The term "sustainable development" was formally introduced in 1987 in the World Commission on Environment and Development's Brundtland Report (see Brundtland, 1987). Since then, many treaties and other instruments have either directly or indirectly supported the notion that states have a responsibility and legal obligation to ensure the sustainable use of natural resources (Kravchenko et al., 2012). The sustainable development norm is reflected in the 1992 Rio Declaration, which contains 27 principles to help guide states toward sustainable development, and the 1992 UN Framework Convention on Climate Change. Although international law lacks a universally accepted definition of sustainable development, it is clear that it is emerging as a principle of international law (Birnie & Boyle, 1992). It is also clear that the spirit of sustainable development, which requires activities to be conducted without causing harm to the environment, is broadly respected by the international community. This is evidenced by global and regional treaties concerning cultural and natural heritage, wildlife conservation, habitat protection, international watercourses, and endangered species, which together imply acceptance of a broader legal significance regarding conservation and sustainable development (Birnie & Boyle, 1992; Kravchenko et al., 2012).

It is worth noting that global environmental governance has not always followed a progressive and linear path, and some have noted a recent "backlash" against such global norms and institutions due to the rise of populism and "illiberal democracy" (Danchin et al., 2020, p. 33). An interesting and important question for both scholars and practitioners alike is whether global environmental norms have informed concrete changes toward sustainability. As the above discussion has highlighted, the application of these twilight norms at the international level is evident, and the examples show how these norms are embedded in various international legal instruments. Yet there is far less clarity surrounding the extent to which these global environmental norms have diffused to the local level. In other words, to what extent do local-level environmental actors draw upon global twilight norms to frame policy and debates, and to influence local action? To help address this question, this research aims to shed light on how these global-level environmental norms are translated and used in specific local contexts. The aim is to help illuminate, at least to some degree, their influence on local-level natural resource governance and management, including how they are invoked to create a "discursive field" (Weedon, 1987) and to frame local policies and actions.

1.3   Framing the Problem

Norms, which refer to "collective expectations for the proper behavior of actors with a given identity" (Katzenstein, 1996, p. 5) are one of the most widely researched areas in contemporary International Relations (IR) scholarship. Norms are essential components of institutions, or the "rules and procedures (both formal and informal) that structure social interaction by constraining and enabling actors' behavior" (Helmke & Levitsky, 2004, p. 727). To understand the impact and relevance of international institutions, there has been much research that examines their influence on the domestic level (see Checkel, 2001; Legro, 1997; Gurowitz, 1999; Farrell, 2001). Although local actors are not ignored in this literature, the perspectives offered on norm diffusion have been criticized for being "unduly static" (Acharya, 2004, p. 243). Following Zwingel (2012), I adopt the perspective that such a static focus "neglects important dynamics of norm adaptation and rejection and ignores that international norms are themselves of evolutionary character" (p. 115). In addition to the international and the national, local dynamics need to be given more attention in order to more fully grasp the significance of international institutions. When it comes to mechanisms of global environmental governance, implementation remains largely within the realm of the nation-state. It follows that ratification of an international environmental treaty, such as those discussed in the previous section, is but one step on the lengthy and complex path toward the realization of environmental benefits, and that local (sub-national) dynamics also represent a critical step on this path.

In theorizing the domestic influence of global twilight norms, I take as the point of departure the literature on norm diffusion. The next chapter reviews this literature and discusses important shortcomings of the diffusion perspective, including its assumption that the state is the most relevant actor when it comes to the implementation of global norms in the domestic context. My focus in this research on instances of local multi-stakeholder governance attempts to shed light on the relevance of a range of actors in processes of norm implementation. In addition, I discuss how the literature on norm diffusion conceptualizes international norms as causal factors that result in an effect in the domestic context – a perspective that fails to take into account processes of negotiation and norm reinterpretation. In critiquing the norm diffusion perspective, I more closely adopt the perspective proposed by Zwingel (2012) that reflects norm *translation*, which centers on context-specific analysis involving multiple actors and contextualized discourses.

Because norms change and evolve through processes of interaction, the research on norm diffusion recognizes that the meaning attached to a particular norm will change to some degree as the norm is translated and applied in different contexts involving different actors (Wiener, 2009, 2016). At the global level, previous research has shown how different sets of stakeholders who represent different identities and interests will often implement global mandates and norms in different ways. For example, Gheciu (2011) has demonstrated how during processes of cooperation, NATO and various NGOs adopted different approaches to peacebuilding and post-conflict reconstruction in ways consistent with their respective organizational cultures. As Checkel (1999, p. 84) highlights, "domestic norms and domestic structure are variables that intervene between systemic norms and national-level outcomes." It follows that when global norms diffuse to the national level, we can expect them to be mediated by context-specific domestic structures and pre-existing domestic norms. Building upon this, Grillot (2011) shows that normative contestation and competition can be expected as international norms are translated to the domestic level. In the domestic context, norm implementation involves different types of stakeholders with a range of identities and interests who may engage in deliberation and contestation processes concerning the legitimacy, applicability, and meaning of particular international norms. Similarly, research has shown how ideational factors in the domestic realm influence and mediate the implementation of global norms, demonstrating how global norms interact with ideational factors at the regional, national, and local levels (Legro, 1997).

In this context of norm diffusion, the literature on norm localization is useful to consider as it explicitly shifts the analytical focus to local-level actors and their active role in norm diffusion processes (Acharya, 2004). For instance, localization can involve processes of *framing* and *grafting* – concepts that will be described in detail in the next chapter – which account for some of the dynamics of norm diffusion and translation (Acharya, 2004). In addition, research on how norms can be deliberately and strategically used to frame issues has examined *foregrounding* and *reframing* processes – two additional concepts that focus on the dynamic aspects of norm translation and diffusion (Raymond et al., 2014). As described in the following chapter, stakeholders may draw upon domestic norms to "foreground," or call attention to and attempt to weaken certain undesirable domestic standards of behavior, and invoke global norms to reframe the issue and propose new norms in their place (Raymond et al., 2014; Raymond, 2016).

In general, research on norms has tended to privilege the international level, while research on norm diffusion and implementation at the domestic level leans toward a focus on national-level political or institutional actors. Research on norm localization has resulted in a much stronger focus on local-level dynamics. This study builds on and contributes to this literature by examining a range of dynamic processes by which global environmental norms are adapted, re-represented, and translated to local contexts. It draws upon the concept of *norm translation* to bring together and examine dynamics that have typically been studied in isolation, including foregrounding and normative reframing on the one hand, and framing and grafting on the other, under a common conceptual framework. This context sets the stage for this study's focus on how global environmental norms are adopted into very localized contexts, and how these norms are used to frame local issues to generate a locally-oriented discourse on environmental sustainability. In addition, by focusing explicitly on environmental norms this study can help address the problem stated by Beyerlin (2008) concerning confusion over the role and effect of global environmental twilight norms. In short, this study aims to contribute to our understanding of how these global environmental norms are adapted and translated to local contexts.

1.4   Local Actors

This research assumes that local actors are important conduits through which global norms are translated into local contexts. The three cases examined in this book, which are described in greater detail in the following sections, each encompass an array of local actors involved in the governance and/or management of one or more local natural resources. In each case, I identify relevant stakeholders, showcase the specific norms drawn upon in discourse surrounding the resource, and highlight how those norms are adapted to the local context; however, I do not explicitly analyze dynamics between or among stakeholders themselves.

In the academic literature, scholarly attention has increasingly focused on efforts to address, measure, monitor, and implement environmental sustainability. Yet much of this attention remains focused on the international and national levels, even though solving environmental sustainability problems requires contributions at various levels of governance, including the local level (Nagy et al., 2018; Kusakabe, 2012). Local actors play a critical role in developing policies, advancing action, and stimulating knowledge to contribute to environmental sustainability goals. In recent years, international actors have promoted international norms pertaining to inclusive and sustainable natural resource governance (Gustafsson et al., 2020), yet we still know very little about how local actors adapt these global norms to specific domestic contexts. In the continually emerging research on norm diffusion, which I discuss in the next chapter, local dynamics tend to be downplayed in favor of a focus on governments, institutions, and transnational civil society. However, local actors often contextualize, adapt, and redefine global or otherwise external norms, suggesting that they are worthy of further study. Prior research on norms has illustrated the marginalized conception of local actors in the diffusion of global norms (Acharya, 2004), and has highlighted their role in redefining and adapting global norms to the local context (Hall, 2013). Local actors, including local civil society and NGOs, have been shown to adopt the role(s) of norm makers, norm takers, norm brokers, and norm entrepreneurs (Sabchev et al., 2021; Hall, 2013) who contextualize, adapt, and localize global or transnational ideas. In general, the role of local actors in translating and localizing global environmental norms is critical to the broad acceptance of those norms and to the success of their expected outcomes for environmental preservation and sustainability. As opposed to being passive recipients of global norms, local actors are active agents of norm localization and can thus help us better understand the continued significance of the global environmental twilight norms discussed above. The following chapter develops this argument in reviewing the literature on norm diffusion and norm localization. I draw on the concepts of *framing* and *grafting* and combine them with insights from the literature on norm-based institutional change to develop a theoretical framework for studying the processes involved in adapting global environmental norms to specific local contexts in France.

1.5   Multi-stakeholder Governance and Environmental Sustainability

This research focuses on three cases of multi-stakeholder governance in assessing how global norms are translated and implemented at the local level. Often called the "gold standard" of private governance (Schleifer, 2016), multi-stakeholder governance involves formal coordination and involvement in decision- and rule-making processes from a diversity of actors, including civil society, the private sector, academic experts, and government (Gleckman, 2018). Multi-stakeholder governance's signature focus is on inclusiveness, and as a result these types of governance arrangements are regarded as having high levels of legitimacy in addressing different policy areas (Fransen, 2012). The global governance literature focuses on multi-stakeholder governance at the global level, whereas there is still little research on multi-stakeholder governance initiatives at the local level.

Overall, the legitimacy of multi-stakeholder governance is commonly considered superior to that of other types of private governance (Fransen, 2012). As a result, global norms that are successfully translated and implemented at the local level through multi-stakeholder initiatives are likely to enjoy high levels of acceptance. Studying these dynamics thus holds implications for how environmental norms can be successfully translated into local contexts and effectively used in framing strategies to help address important sustainability issues not only in France, but also elsewhere.

While the main contribution of this study lies in generating deeper insights into the dynamics of how global norms are applied in local settings, the specific issue area of environmental sustainability and environmental conservation on which I focus provides a relevant context in which to explore the research questions. The area of environmental sustainability/conservation offers a particularly rich opportunity to identify cases of multi-stakeholder governance at the local level. Because of their high level of importance, environmental issues have become a popular area in which to observe multi-stakeholder governance. As others have noted,

"environmental governance is of growing concern in a world that is more intercon-
nected and interdependent than ever and threatened by climate change and the
depletion or contamination of natural resources that are often shared by multiple
stakeholders" (Gluesing et al., 2017, p. 211). Simply put, there are perhaps more
examples of local multi-stakeholder initiatives to observe in the issue area of the
environment compared to other global issues. One reason for this is because of the
influence of the United Nations' global sustainable development agenda in promoting multi-stakeholder arrangements at different levels of governance for addressing
sustainable development challenges (Reed, 2008). Examples of these efforts include
Agenda 21, which outlines an action plan for sustainable development at the global,
national, and local levels, as well as Local Agenda 21 (LA21) and Sustainable
Development Goal (SDG) #17: Partnerships for the SDGs (Macdonald et al., 2019).

The country context of France, which is further considered in the following section, provides numerous examples of local multi-stakeholder governance involving
environmental sustainability and conservation for researchers to observe. France
thus provides a promising "test case" for investigating the research questions put
forth in this study.

1.6   France as a Backdrop

France was selected according to the logic of a most-likely case design as the country setting in which three local cases will be observed. The main consideration for
examining France centers on its strong national focus on environmental conservation, preservation, and sustainability – in other words, we can observe a high degree
of congruence between international norms that focus on environmental sustainability and preservation, and domestic environmental norms in France. Most-likely
case designs "are based on the assumption that some cases are more important than
others for the purposes of testing a theory" (Levy, 2008, p. 12). Ostrom (1990) suc-
cinctly explains this logic in defending her choice of cases in her Nobel Prize win-
ning research when she draws parallels to how biologists conduct empirical
observation. She writes, "The organism is not chosen because it is representative of
all organisms. Rather, the organism is chosen because particular processes can be
studied more effectively using this organism than using another" (Ostrom, 1990,
p. 26). It is this logic employed in selecting France as the "organism" within which
to observe examples of local multi-stakeholder governance on environmental issues.

Compared to other advanced democracies, France is a leader when it comes to
environmental governance. It has an ambitious and well-developed environmental
policy agenda, which is "aimed chiefly at cutting greenhouse gas (GHG) emissions
but also at dealing with local air and water pollution, waste management and the
conservation of biodiversity" (Egert, 2012, p. 1). France was also the leader in
launching the Global Pact for the Environment with the United Nations General
Assembly (Human Rights Watch, 2017), demonstrating a unique commitment
among the world's nation-states to environmental preservation and conservation.

The strong emphasis in France on environmental conservation, preservation, and sustainability makes it a highly likely setting in which to observe global environmental twilight norms being adapted in different local contexts. Social movement researchers would describe this institutional setting as being characterized by an open political opportunity structure (Tarrow, 1994); in other words, it offers a favorable political and institutional environment for environmental sustainability and conservation work.

From a theoretical perspective, such a setting can potentially shed light on how norm adaptation processes unfold when congruence is high between international norms and domestic norms. As I discuss in the next chapter, the literature examining norm localization adopts a strong focus on settings characterized by low levels of initial congruence between international and domestic norms. Although such a focus has been useful in enabling scholars to pinpoint congruence-building dynamics and processes, I argue that these are not the only settings in which norm adaptation processes unfold. In the following chapter I discuss why we should also expect to observe adaptation dynamics in cases where congruence between international norms and domestic norms is rather high, as in the case of France, and why such a focus can generate deeper theoretical insights into how global norms are applied at the local level.

1.7 Research Design: Examining Three Local-Level Cases

Within France, the three cases that I examine are: the Cerbère-Banyuls Marine Nature Reserve, the Thau Fisheries Local Action Group, and the Biovallée biodistrict. Each case is described in detail below.

1.7.1 The Cerbère-Banyuls Marine Nature Reserve

Founded in 1974, the Cerbère-Banyuls Marine Nature Reserve was the first natural marine reserve in France and is thus also the oldest (Marine Conservation Institute, 2020). In the early 1970s, the local community came together, negotiated, and formalized basic management rules to maintain and preserve the marine area over time (Le Département des Pyrénées-Orientales, 2020). They perceived that formal management and rules were needed because of degradation of the marine environment due to increasing levels of tourism (Le Département des Pyrénées-Orientales, 2020). Today, the local community is still involved in managing the reserve, although it is formally under surveillance by the General Council of the Pyrénées-Orientales.

The marine reserve occupies 6.5 km of coast and extends roughly 1.5 km to the sea (Marine Conservation Institute, 2020). It has two levels of protection: the first involves a protected area corresponding to the largest area of the reserve, which is approximately 600 ha where human activities are regulated, and the second involves a protected area of 65 ha where human activities are strictly prohibited (Marine Conservation Institute, 2020). Each year, the marine reserve carries out surveillance missions, scientific monitoring, educational activities, and manages the reception of users and the public. The stakeholders involved in governing and managing the marine reserve include scientific experts, local community members, and the French government.

1.7.2 The Thau Fisheries Local Action Group (FLAG)

FLAGs are partnerships between fisheries actors and other private actors, local authorities, and civil society organizations. These stakeholders collectively design and implement a local development strategy to address their area's economic, social, and/or environmental needs (European Commission, 2020a). The FLAG is somewhat different than the other two cases in that it selects and provides funding to local projects that contribute to local development in the area. In general, FLAGs also involve the European Union (EU) level in that they are funded by the European Maritime and Fisheries Fund (EMFF), promoting community-led local development (CLLD) (European Commission, 2020c).

The Thau FLAG covers an area along the Mediterranean from Frontignan, France to Agde, France, with two important fishing harbors in Agde and Sète. The FLAG's strategy involves the promotion of local economic activities while considering and weighing heavily the unique environmental assets and challenges of the area (European Commission, 2020b). The specific related challenges consist of creating and maintaining jobs and businesses in the fishery sector, and strengthening the importance of fishery and aquaculture activities within territorial development while at the same time ensuring sustainable development. This particular area is experiencing several environmental challenges involving water quality and sustainable management of land use, which have been exacerbated by the increase in tourism over the past decade.

The Thau FLAG involves 75 individuals and organizations working in partnership, which includes a mix of stakeholders such as fisheries actors, public authorities, NGOs, and environmental experts (European Commission, 2020b). The Thau FLAG has stakeholders based in Sète, France and in the relevant EU office in Brussels, Belgium.

1.7.3 The Biovallée Biodistrict

Biodistricts are territories "where farmers, citizens, associations and public authorities enter into an agreement for the sustainable management of local resources" and are considered participatory and innovative multi-stakeholder governance arrangements (Assaël, 2017, p. 1.). As the International Network of Eco-Regions (n.d, p. 2) explains:

> The idea behind the bio-district approach is to create and reinforce links that would benefit everyone involved: organic farmers would get better market access, consumers would benefit from transparency about the origins of their food and enjoy fresh, organically grown local products, the tourist operators would offer new activities and destinations (eco trails and agro-tourism farms) while public authorities would ensure food security and rural employment. The multifunctional approach is the very innovation. It allows to incorporate various fields of agriculture combined with other farming activities: eco-tourism, education, culture, leisure, landscape preservation.

Founded in 2009, the Biovallée biodistrict is the only biodistrict in France, consisting of an association of municipalities in the Drôme Valley, located in the Rhône-Alpes region. Encompassing 54,000 inhabitants in 102 small towns and villages, the biodistrict has "undergone a paradigm shift towards organic agriculture over the past few decades," whereby "40% of the valley's farmers use organic practices, compared to just 8% nationwide" (Local Futures, 2020, n.p.). The Biovallée biodistrict aims to "steward a region-wide sustainable development masterplan that will serve as a blueprint for the rest of France" (Local Futures, 2020, n.p.). There are multiple stakeholders in the Biovallée project who coordinate efforts in a multi-stakeholder governance framework at the local level "to plan and implement alternative energy production, conversion to organic agriculture, waste reduction, local food procurement for institutions, land use planning to slow urbanization, and more" (Local Futures, 2020, n.p.). The stakeholders include elected officials, farmers, entrepreneurs, and civil society.

The selection of these three particular cases follows a straightforward logic. For one, the three cases are all observed in the same country context, allowing the national-level institutional setting to be held constant. Furthermore, these cases share many other similarities (e.g., issue scope, a focus on local-level action) and are thus broadly comparable. In addition, these cases are similar in that each reflects a local instance of multi-stakeholder governance involving environmental conservation and/or sustainability. Although each case involves multi-stakeholder governance, there is variation in what form it takes, how the governance institutions are designed, and how participatory they are. This variation lends itself to a comparison across cases of the dynamics involved in adapting global norms to different local settings.

This research design combines within-case analysis with cross-case comparisons (George & Bennett, 2005). Thus, it uses a comparative, multiple-case study research design to study how global environmental norms are adapted and translated to the sub-national level. Multiple case studies enable the researcher to examine the same phenomenon across different cases. As others have observed, "it is somewhat analogous to the concept of 'replication' in group (or nomothetic) designs. Thus, if an investigator wants to repeat a study to strengthen theory or test the findings of a single case on other cases, a multiple-case study approach is preferred" (DePoy & Gitlin, 2016, p. 174). In this project, the aim is to investigate the same phenomenon across different micro-level settings in order to build theory and collect empirical evidence on the dynamic aspects of diffusion of global environmental norms to the local level. This type of design allows for an in-depth analysis of those processes (Geertz, 1973).

1.8 Materials and Methodology

Identifying a norm can be a difficult task but is nonetheless possible because "norms by definition embody a quality of 'oughtness' and shared moral assessment," and they "prompt justifications for action and leave an extensive trail of communication among actors that we can study" (Finnemore & Kathryn, 1998, p. 892; see also Goertz & Diehl, 1992). The documents that stakeholders in these local settings produce in relation to their natural resource governance and management activities present excellent opportunities to examine and analyze this "trail of communication" for the presence and usage of global environmental norms. The analysis is thus based primarily on a comprehensive examination of hundreds of pages of publicly available documents and relevant website materials specific to each case. The websites related to each case hold much potential as resources, as they contain a rich body of information about each case, its stakeholders, and its philosophy, values, history, activities, and practices. All sources were analyzed in their original language (French). This publicly available information presents an opportunity to examine how global environmental norms are adapted, translated, and contextualized in each setting.

Accordingly, I use directed content analysis as the primary methodology. Unlike conventional content analysis in which researchers avoid using preconceived categories (Kondracki et al., 2002), directed content analysis adopts a more structured process. It involves using existing theory or prior research to identify key concepts or categories, and to settle on operational definitions prior to the analysis (Potter & Levine-Donnerstein, 1999). It can be thought of as deductive category application (Mayring, 2000). As discussed earlier, this research examines the adaptation of so-called environmental twilight norms to the local level. The directed content analysis thus primarily involves a deductive approach to identifying the extent to which these twilight norms are evident in the three cases examined.

Using the literature on localization and norm-based institutional change as a guide, I employ the following coding scheme to capture dynamic processes: (1) twilight norms that are promoted (norm identification and usage), (1a) How a twilight norm is used to frame a local issue (framing), (1b) How a twilight norm is associated with another preexisting norm in the same issue space (grafting), (2) preexisting norms that are criticized (foregrounding: bringing an existing norm to the surface in order to weaken it), (2a) whether or not a criticized norm's "fit" to the issue is called into question (normative reframing), and (2b) whether or not the content or character of a criticized norm itself is rejected (normative innovation). This coding process enables the study of global twilight norms in local contexts while capturing concepts from the literatures on norm localization and norm-based institutional change that highlight dynamic processes of diffusion driven by local actors.

In addition to the document analysis, participant observation was used as a secondary research methodology. Visits were conducted at the case settings in France during the month of July 2022, during which time I observed stakeholders involved in the day-to-day operation and management of the specific case, participated in on site activities, and/or engaged in informal conversations on site that were recorded as field notes to supplement the content analysis.

1.9   Contributions of the Research

This study makes several contributions to the literature. First, it applies the theoretical framework developed in the next chapter in ways that deepen our understanding of the ways in which global environmental norms are translated and applied in local settings. By combining concepts from the literature on norm localization with processes from the literature on norm-based institutional change, this study can potentially generate new insights in explicitly focusing on and highlighting the processes of norm translation across three different localized contexts within the same country. Additionally, in focusing the analysis on twilight norms, this study can help shed light on the continued significance of these global environmental norms, helping to address Beyerlin's (2008) observation regarding the confusion over the role that these twilight norms play in environmental governance and the effect that they have.

Finally, the emphasis of this research on multi-stakeholder governance opens up a new avenue for examining norm translation in new cases involving innovative governance practices. In addition to contributing a comparative case study to the literature on localization and norm diffusion, this focus may help us better appreciate the complex array of local actors involved in adapting global norms to local contexts and develop more nuanced accounts of broader efforts to bring about environmental change on different scales and in different contexts. In helping to move the focus beyond the nation-state in understanding how global norms are implemented within a specific domestic context, this research contributes to the understudied area of local-level dynamics in norm implementation processes.

1.10   Plan of the Book

The remainder of the book is organized into several chapters. Chap. 2 situates this research in the theoretical context of norm diffusion. It reviews the literatures on norm localization and norm-based institutional change, bridging concepts from these literatures that focus on dynamic aspects of norm diffusion driven by local actors. From there, the book is organized around three case studies in France that examine the use of specific global environmental twilight norms and how they are being adapted to the local level. Chap. 3 examines the case of the Cerbère-Banyuls Marine Nature Reserve, Chap. 4 focuses on the Thau Fisheries Local Action Group, and Chap. 5 analyzes the Biovallée biodistrict. Each of these empirical chapters identifies the global twilight norms that are evident and assesses their use in the respective local context, focusing on the research questions put forth at the beginning of this chapter. Finally, the Conclusion summarizes the main findings of the study and relates them to the relevant scholarship, draws comparisons across cases, and discusses avenues for future research.

References

Acharya, A. (2004). How ideas spread: Whose norms matter? Norm localization and institutional change in Asian regionalism. *International Organization, 58*(2), 239–275. https://doi.org/10.1017/S0020818304582024

Akintoba, T. O. (1996). *African states and contemporary international law: A case study of the 1982 law of the sea convention and the exclusive economic zone.* Martinus Nijhoff Publishers.

Assaël, K. (2017). *Creating a system of bio-districts in Italy within the national policies.* Retrieved from https://kipschool.org/usr_files/generic_pdf/BiodistrictsItalia2017-ENG.pdf

Barral, V. (2012). Sustainable development in international law: Nature and operation of an evolutive legal norm. *European Journal of International Law, 23*(2), 377–400. https://doi.org/10.1093/ejil/chs016

Baslar, K. (1998). *The concept of the common heritage of mankind in international law.* Martinus Nijhoff.

Beyerlin, U. (2008). Different types of norms in international environmental law: Policies, principles, and rules. In D. Bodansky, J. Brunnée, & E. Hey (Eds.), *The Oxford handbook of international environmental law* (pp. 425–448). Oxford University Press.

Bilder, R. B. (1980). International law and natural resources policies. *Natural Resources Journal, 20*(3), 451–486. Retrieved from https://digitalrepository.unm.edu/nrj/vol20/iss3/3

Birnie, P. W., & Boyle, A. E. (1992). *International law and the environment.* Clarendon Press.

Brundtland, G. (1987). Report of the world commission on environment and development: Our common future. United Nations General Assembly document A/42/427.

Checkel, J. T. (1999). Norms, institutions, and national identity in contemporary Europe. *International Studies Quarterly, 43*(1), 83–114. https://doi.org/10.1111/0020-8833.00112

Checkel, J. T. (2001). Why comply? Social learning and European identity change. *International Organization, 55*(3), 553–588. https://doi.org/10.1162/00208180152507551

Danchin, P. G., Farrall, J., Ford, J., Rana, S., Saunders, I., & Verhoeven, D. (2020). Navigating the backlash against global law and institutions. *The Australian Year Book of International Law Online, 38*(1), 33–77. https://doi.org/10.1163/26660229_03801004

DePoy, E., & Gitlin, L. N. (2016). *Introduction to research (fifth edition): Understanding and applying multiple strategies.* Elsevier.

Egert, B. (2012). *France's environmental policies: Internalising global and local externalities* (CESifo working paper series 3887). CESifo. Retrieved from https://www.cesifo.org/en/publikationen/2012/working-paper/frances-environmental-policies-internalising-global-and-local

European Commission. (2020a). *FLAG factsheets.* Retrieved from https://webgate.ec.europa.eu/fpfis/cms/farnet2/on-the-ground/flag-factsheets-list_en

European Commission. (2020b). *FLAG factsheet: Thau FLAG.* Retrieved from https://webgate.ec.europa.eu/fpfis/cms/farnet2/on-the-ground/flag-factsheets/thau-flag_en

European Commission. (2020c). *FARNET – The European fisheries areas network*. Retrieved from https://ec.europa.eu/fisheries/cfp/eff/farnet_en

FADEAR. (n.d.-a). *Notre histoire*. Retrieved from https://www.agriculturepaysanne.org/Notre-histoire. Accessed 14 June 2021.

FADEAR. (n.d.-b). *L'agriculture paysanne*. Retrieved from https://www.agriculturepaysanne.org/IMG/pdf/plaquette_10principes_off_bd-2.pdf. Accessed 14 June 2021.

FADEAR. (n.d.-c). *Charte de l'agriculture paysanne*. Retrieved from https://www.agriculturepaysanne.org/IMG/pdf/charte-agriculture-paysanne.pdf. Accessed 10 June 2021.

FADEAR. (2021). Les 10 principes politiques de l'Agriculture paysanne. Retrieved from https://www.agriculturepaysanne.org/Les-10-principes-politiques-de-l-Agriculture-paysanne. Accessed 12 June 2021.

Farrell, T. (2001). Transnational norms and military development: Constructing Ireland's professional army. *European Journal of International Relations, 7*(1), 63–102. https://doi.org/10.1177/1354066101007001003

Finnemore, M., & Kathryn, S. (1998). International norm dynamics and political change. *International Organization, 52*(4), 887–917. https://doi.org/10.1162/002081898550789

Fransen, L. (2012). Multi-stakeholder governance and voluntary programme interactions: Legitimation politics in the institutional design of corporate social responsibility. *Socio-Economic Review, 10*(1), 163–192. https://doi.org/10.1093/ser/mwr029

Geertz, C. (1973). *The interpretation of cultures: Selected essays*. Basic Books.

George, A. L., & Bennett, A. (2005). *Case studies and theory development in the social sciences*. MIT Press.

Gheciu, A. (2011). Divided partners: The challenges of NATO-NGO cooperation in peacebuilding operations. *Global Governance, 17*(1), 95–113. https://doi.org/10.1163/19426720-01701007

Gleckman, H. (2018). *Multistakeholder governance and democracy: A global challenge*. Routledge.

Gluesing, J., Riopelle, K., & Wasson, C. (2017). Environmental governance in multi-stakeholder contexts: An integrated methods set for examining decision-making. In B. Hollstein, W. Matiaske, & K.-U. Schnapp (Eds.), *Networked governance: New research perspectives* (pp. 211–244). Springer.

Goertz, G., & Diehl, P. F. (1992). Toward a theory of international norms: Some conceptual and measurement issues. *Journal of Conflict Resolution, 36*(4), 634–664. https://doi.org/10.1177/0022002792036004002

Grillot, S. R. (2011). Global gun control: Examining the consequences of competing international norms. *Global Governance, 17*(4), 529–555. https://doi.org/10.1163/19426720-01704008

Gurowitz, A. (1999). Mobilizing international norms: Domestic actors, immigrants, and the Japanese state. *World Politics, 51*(3), 413–445. https://doi.org/10.1017/S0043887100009138

Gustafsson, M.-T., Merino, R., & Scurrah, M. (2020). Domestication of international norms for sustainable resource governance: Elite capture in Peru. *Environmental Policy and Governance, 30*(5), 227–238. https://doi.org/10.1002/eet.1904

Hall, R. B. (2013). *Reducing armed violence with NGO governance*. Routledge.

Helmke, G., & Levitsky, S. (2004). Informal institutions and comparative politics: A research agenda. *Perspectives on Politics, 2*(4), 725–740. https://doi.org/10.1017/S1537592704040472

Human Rights Watch. (2017). *France jumpstarts initiative to protect environment*. 19 September. Retrieved from https://www.hrw.org/news/2017/09/19/france-jumpstarts-initiative-protect-environment#

International Convention for the Regulation of Whaling, Dec. 2, 1946.62 Stat. 1716; 161 UNTS 72.

International Network of Eco-Regions -IN.N.E.R Association. (n.d.). *52 profiles on agroecology: The experience of bio-districts in Italy*. Retrieved from http://www.fao.org/3/a-bt402e.pdf

Katzenstein, P. (1996). *The culture of National Security: Norms and identity in world politics*. Columbia University Press.

Kondracki, N. L., Wellman, N. S., & Amundson, D. R. (2002). Content analysis: Review of methods and their applications in nutrition education. *Journal of Nutrition Education and Behavior, 34*(4), 224–230. https://doi.org/10.1016/S1499-4046(06)60097-3

Kotzé, L. J., & Muzangaza, W. (2018). Constitutional international environmental law for the Anthropocene? *Review of European, Comparative, & International Environmental Law, 27*(3), 278–292. https://doi.org/10.1111/reel.12244

Kravchenko, S., Chowdhury, T. M. R., & Bhuiyan, J. H. (2012). Principles of international environmental law. In S. Alam, J. H. Bhuiyan, T. M. R. Chowdhury, et al. (Eds.), *Routledge handbook of international environmental law* (pp. 43–60). Routledge.

Kusakabe, E. (2012). Social capital networks for achieving sustainable development. *Local Environment, 17*(10), 1043–1062. https://doi.org/10.1080/13549839.2012.714756

Le Département des Pyrénées-Orientales. (2020). *Découvrir l'histoire de la Réserve Naturelle Marine de Cerbère-Banyuls*. Retrieved from https://www.ledepartement66.fr/decouvrir-lhistoire-de-la-reserve-naturelle-marine-de-cerbere-banyuls/

Legro, J. W. (1997). Which norms matter? Revisiting the "failure" of internationalism. *International Organization, 51*(1), 31–63. https://doi.org/10.1162/002081897550294

Levy, J. S. (2008). Case studies: Types, designs, and logics of inference. *Conflict Management and Peace Science, 25*(1), 1–18. https://doi.org/10.1080/07388940701860318

Local Futures. (2020). *Biovallée project: France*. Retrieved from https://www.localfutures.org/programs/global-to-local/planet-local/food-farming-fisheries/biovallee-project/

Macdonald, A., Clarke, A., & Huang, L. (2019). Multi-stakeholder partnerships for sustainability: Designing decision-making processes for partnership capacity. *Journal of Business Ethics, 160*(3), 409–426. https://doi.org/10.1007/s10551-018-3885-3

Marine Conservation Institute. (2020). *Cerbère – Banyuls – partially protected zone*. Marine Protection Atlas. Retrieved from https://mpatlas.org/zones/414/

Mayring, P. (2000). Qualitative content analysis. *Forum: Qualitative Social Research, 1*(2). Retrieved from http://www.qualitative-research.net/fqs-texte/2-00/02-00mayring-e.htm

McIntyre, O., & Mosedale, T. (1997). The precautionary principle as a norm customary international law. *Journal of Environmental Law, 9*(2), 221–241. https://doi.org/10.1093/jel/9.2.221

Nagy, J. A., Benedek, J., & Ivan, K. (2018). Measuring sustainable development goals at a local level: A case of a metropolitan area in Romania. *Sustainability, 10*(11), 3962. https://doi.org/10.3390/su10113962

Nanodkar, S. (2018). Polluter pays principle: Essential element of environmental law and policy. *International Journal of Law Management & Humanities, 1*(5), 1–8. Retrieved from https://www.ijlmh.com/wp-content/uploads/2019/03/Polluter-Pays-Principle-Essential-Element-of-Environmental-Law-and-Policy.pdf

OECD. (1972). Council recommendation on guiding principles concerning the international economic aspects of environmental policies of the Organisation for Economic Co-operation and Development, C(72), 128.

Ostrom, E. (1990). *Governing the commons: The evolution of institutions for collective action*. Cambridge University Press.

Potter, W. J., & Levine-Donnerstein, D. (1999). Rethinking validity and reliability in content analysis. *Journal of Applied Communication Research, 27*(3), 258–284. https://doi.org/10.1080/00909889909365539

Qureshi, W. A. (2019). Protecting the common heritage of mankind beyond national jurisdiction. *Arizona Journal of International & Comparative Law, 36*(1), 82–109. Retrieved from http://arizonajournal.org/wp-content/uploads/2019/04/Qureshi_36.1_.pdf

Raymond, L. (2016). *Reclaiming the atmospheric commons: The regional greenhouse gas initiative and a new model of emissions trading*. MIT Press.

Raymond, L., Weldon, S. L., Kelly, D., Arriaga, X. B., & Clark, A. M. (2014). Making change: Norm-based strategies for institutional change to address intractable problems. *Political Research Quarterly, 67*(1), 197–211. https://doi.org/10.1177/1065912913510786

Reed, M. (2008). Stakeholder participation for environmental management: A literature review. *Biological Conservation, 141*(10), 2417–2431. https://doi.org/10.1016/j.biocon.2008.07.014

Rosencranz, A. (2003). The origin and emergence of international environmental norms. *Hastings International and Comparative Law Review, 26*(3), 309–320. Retrieved from https://repository.uchastings.edu/cgi/viewcontent.cgi?article=1580&context=hastings_international_comparative_law_review

Sabchev, T., Miellet, S., & Durmus, E. (2021). Human rights localization and individual agency: From "hobby of the few" to the few behind the hobby. In C. Boost, A. Broderick, F. Coomans, & R. Moerland (Eds.), *Myth or lived reality: On the (in)effectiveness of human rights* (pp. 183–212). Asser Press.

Schleifer, P. (2016). Let's bargain: Setting standards for sustainable biofuels. In A. Esguerra, N. Helmerich, & T. Risse (Eds.), *Sustainability politics and limited statehood: Contesting the new modes of governance* (pp. 47–73). Springer.

Stone, C. D. (2004). Common but differentiated responsibilities in international law. *The American Journal of International Law, 98*(2), 276–301. https://doi.org/10.2307/3176729

Tanzi, A. M. (2020). The inter-relationship between no harm, equitable and reasonable utilisation and cooperation under international water law. *International Environmental Agreements: Politics, Law and Economics, 20*, 619–629. https://doi.org/10.1007/s10784-020-09502-7

Tarrow, S. (1994). *Power in movement: Social movements, collective action and politics.* Cambridge University Press.

Taylor, P. (2019). The common heritage of mankind: Expanding the oceanic circle. In International Ocean institute – Canada (Ed.), *The future of ocean governance and capacity development* (pp. 142–150). Brill | Nijhoff.

United Nations. (1973). Declaration of the United Nations conference on the human environment, 16 June 1972, U.N. Doc. A/Conf.48/14/Rev. 1, reprinted in 11 ILM 1416. Retrieved from https://www.refworld.org/docid/3b00f1c840.html

United Nations. (1990). *Bergen ministerial declaration on sustainable development in the ECE region: Sustainable development meets the needs of the present without compromising the ability of the future generations to meet their own needs.* United Nations, Economic Commission for Europe.

Wang, T., & Gao, X. (2018). Reflection and operationalization of the common but differentiated responsibilities and respective capabilities principle in the transparency framework under the international climate change regime. *Advances in Climate Change Research, 9*(4), 253–263. https://doi.org/10.1016/j.accre.2018.12.004

Weedon, C. (1987). *Feminist practice and poststructuralist theory.* Blackwell.

Weiss, E. B. (1996). *In fairness to future generations: International law, common patrimony, and intergenerational equity.* Transnational Publishers.

Wiener, A. (2009). Enacting meaning-in-use: Qualitative research on norms and international relations. *Review of International Studies, 35*(1), 175–193. https://doi.org/10.1017/S0260210509008377

Wiener, A. (2016). Contested norms in inter-national encounters: The 'Turbot War' as a prelude to fairer fisheries governance. *Politics and Governance, 4*(3), 20–36. https://doi.org/10.17645/pag.v4i3.564

Yamin, F., & Depledge, J. (2004). *The international climate change regime: A guide to rules, institutions and procedures.* Cambridge University Press.

Zwingel, S. (2012). How do norms travel? Theorizing international women's rights in transnational perspective. *International Studies Quarterly, 56*(1), 115–129. https://doi.org/10.1111/j.1468-2478.2011.00701.x

Chapter 2
How Global Norms Travel: A Theoretical Framework

This chapter outlines a theoretical framework for analyzing how global environmental norms are adapted and applied at the local level. It begins by reviewing the literature on norm diffusion and articulating some of the critiques that the diffusion concept has been subject to. I build on this by subsequently introducing the literatures on norm localization and norm-based institutional change. I emphasize their focus on dynamic concepts, such as framing, grafting, and foregrounding, and relate them to processes of frame construction – an under-researched aspect of issue framing. Lastly, I review literature on actor constellations and highlight the concept of impact translation in order to emphasize the theoretical relevance of non-state actors and embedded context in processes of norm translation and diffusion. Overall, the aim of this chapter is to develop a theoretical framework that can be used to investigate how global norms travel to the local level and are applied by stakeholders involved in the cases of environmental governance and natural resource management examined in the subsequent chapters.

2.1 Norm Diffusion

Scholars who study norms have developed an extensive theoretical and empirical literature addressing the emergence, diffusion, and effect of norms in the international system. Specifically, the literature on norm diffusion is concerned with how norms travel – that is, how they move from their original context and are translated into new contexts (Winston, 2018). At the international level, norms travel horizontally across states in the international system. Theories of norm diffusion explain how norms gain influence and impact states' identities and behavior (Finnemore & Sikkink, 1998). Scholars have identified several mechanisms of international norm diffusion, including coercion (Hafner-Burton, 2005), competition (Boehmke & Witmer, 2004), emulation (Elkins & Simmons, 2005), and learning (DiMaggio & Powell, 1983).

© The Author(s), under exclusive license to Springer Nature
Switzerland AG 2023
M. Schnyder, *Global Norms in Local Contexts*, SpringerBriefs in Political
Science, https://doi.org/10.1007/978-3-031-41108-3_2

In general, an international norm starts with an innovative idea developed by norm entrepreneurs – those who mobilize support to create alternative ideas (Sunstein, 1996) – and ends as a far-reaching principle that becomes widely institutionalized among states. To capture what happens in between this beginning and end state, the norm life cycle model developed by Finnemore and Sikkink (1998) explains how norms evolve and diffuse internationally. The "life cycle" according to which norms spread consists of three stages: norm emergence, norm cascade, and norm internalization (Finnemore & Sikkink, 1998). The first stage, norm emergence, is characterized by the promotion of a new norm by norm entrepreneurs who attempt to persuade states to adopt it. When a critical mass of states has accepted a norm, a tipping point is reached. This then sets the stage for the second stage of the norm life cycle, norm cascade, as early adopters of the norm alter the context for the remaining adopters and influence other states to adopt the emerging norm. During this period of norm cascade, the new norm begins to enjoy wide acceptance. The third stage, norm internalization, takes root when compliance with the new norm becomes almost automatic and the norm's validity is no longer debated (Finnemore & Sikkink, 1998).

Although the norm life cycle model has produced important insights into the process of norm evolution, it has been critiqued for its static conception of norm institutionalization (Welsh, 2013; Payne, 2001). For instance, the life cycle model does not consider that institutionalized norms could be interpreted differently by different actors, contested, and reinterpreted and redefined over time. In short, the linear process that the life cycle model describes assumes that once a norm has been institutionalized, it remains stable and its meaning does not change or evolve (Welsh, 2013). Moreover, in focusing on transnational agents at the level of the international system, the life cycle model leaves the "domestic dynamics of norm creation and appropriation underestimated" (Zwingel, 2012, p. 118).

By adopting a strong focus on domestic agency and process, the "second wave" scholarship on norm diffusion attempts to remedy this weakness (Cortell & Davis, 2000). In this strand of the literature, the role of domestic cultural, organizational, and political variables in conditioning or mediating the diffusion of global norms is emphasized (Risse-Kappen, 1994; Cortell & Davis, 1996; Legro, 1997; Checkel, 1999, 2001). For example, Cortell and Davis (2000) analyze the impact (salience) of international norms within the domestic environment, examining their influence on state policies and institutions, as well as national discourse. Their findings point to the relevance of pre-existing domestic discourse, which "provides the context within which the international norm takes on meaning" (Cortell & Davis 2000, p. 73). This idea reflects the concept of congruence, which describes the compatibility between specific international norms and domestic-level norms. The congruence concept holds particular importance for assessing the likelihood of norm diffusion from the international level to the domestic environment. Similarly, Checkel (1999) develops the idea of "cultural match," which occurs when an international norm's prescriptions converge with a state's domestic norms, in arguing that norm diffusion is facilitated when an international norm resonates with pre-existing domestic norms. Conversely, Grillot posits that "the clash between domestic and international norms may prevent or affect the diffusion and internalization of new ideas" (2011, 534). In focusing on the notion of organizational culture, which also reflects the idea of congruence, Legro (1997) demonstrates how domestic-level ideational factors condition the diffusion of global norms. Savery (2007) also highlights the notion of congruence in analyzing why international sexual non-discrimination norms have diffused unevenly across the international system, arguing that one must carefully examine the facilitating and impeding factors at the domestic level in terms of how they relate to a specific international norm. In short, this stream of literature has demonstrated that the implementation of international norms at the domestic level critically depends on the degree of congruence between the norm and the prevailing attitudes of domestic actors (Botcheva & Martin, 2001; Checkel, 1999; Young, 1999), and the overall compatibility of an international norm with the material interests of a state (Cardenas, 2007).

Other studies have critiqued this perspective of congruence for being "unduly static," (Acharya, 2004, p. 243) in simply conceptualizing whether or not a "match" exists between international and domestic norms. This line of research has sought to refine the concept by taking account of the dynamic aspects that the congruence process entails in investigating the complex processes by which congruence-building unfolds in the domestic arena. In other words, it moves away from viewing congruence in terms of a static fit between norms and toward the view that congruence entails dynamic acts of congruence-building. Such acts are made possible in the first place because "internationally binding norms are frequently characterized by a certain degree of ambiguity," and this ambiguity "leaves space for interpretation" in translating global norms to the domestic context (Eimer et al., 2016, p. 451; see also Kersbergen & Verbeek, 2007). Domestic actors can utilize this "space for interpretation" to integrate their own preferences.

Acharya (2004) helps illustrate this idea in developing the notion of localization, which he defines as "the active construction (through discourse, framing, grafting, and cultural selection) of foreign ideas by local actors, which results in the former developing significant congruence with local beliefs and practices" (p. 245). The localization perspective is additionally useful for broadening the set of actors considered relevant norm entrepreneurs. In essence, it redefines the way we understand norm entrepreneurship by shifting the focus away "from 'outsider proponents' committed to a transnational or universal moral agenda to 'insider proponents,'" which can be "individuals, regionally based epistemic communities, or nongovernmental organizations (NGOs), whose primary commitment is to localize a normative order and whose main task is to legitimize and enhance that order by building congruence with outside ideas" (Acharya, 2004, p. 249). In other words, local stakeholders play the central role. Research examining norms for sustainable and inclusive resource governance has adopted this perspective in showing how national elites and societal coalitions institutionalize these norms in the domestic context very differently (Gustafsson & Scurrah, 2019). Adopting this dynamic perspective of congruence ultimately helps to shed light on how existing domestic norms can redefine an international norm in a specific local context.

2.2   Localization Dynamics

The process of translating an international norm to a specific domestic context entails a range of dynamic, adaptive acts performed by local stakeholders. This section highlights and explains some of those tactics. It draws from the literatures on localization and norm-based framing, and combines dynamic acts identified in these literatures under an overarching theoretical framework for investigating how global-level environmental norms are adapted and translated to specific local contexts.

The localization literature tends to focus on cases in which there is a low degree of initial congruence between international and domestic norms. Such a focus makes sense, as it can facilitate the study of congruence-building and the dynamics that it entails. However, I argue that these same dynamics can also take place in cases where congruence between international norms and domestic norms is rather high, as in the country context of France investigated here. This argument assumes that "a full-fledged congruence between an international norm and the domestic constellation of public and private preferences can be expected to be rather an exception than the rule" (Eimer et al., 2016, p. 454); in other words, pure norm adoption in which "an international norm is implemented in the domestic context without any significant modifications" (Eimer et al., 2016, p. 455) can be regarded as a very rare occurrence. This assumption opens the expectation of observing processes of norm adaptation and translation across a broad range of cases, even when initial levels of congruence may be high (but not perfect). This is because, as discussed earlier, international norms are characterized by a degree of ambiguity that allows domestic actors to bring in their own preferences and perspectives in translating them to the local setting. Different interests and perspectives can remake norms by changing their content, or their clarity and specificity (Sandholtz, 2008). Furthermore, even in cases where congruence is high, it is possible for different local stakeholders to have varying interpretations of a given norm, since norm implementation by different actors operating in specific contexts can result in competing meanings. Although international norms may have formal and legal validity, they require social recognition in order to be translated to specific environments. In short, the meaning of a norm is "enacted at a particular point in time by particular actors" (Wiener, 2009, p. 180), which gives rise to the possibility of competing interpretations. The ambiguity inherent in international norms thus paves the way for norm translation to the local context to entail a range of dynamic processes, which are discussed next.

Acharya (2004) argues that the concept of *grafting* is important to recognize in studying the translation of a global norm to a domestic setting in that it offers a dynamic lens through which to view congruence processes. Grafting, which has also been called incremental norm transplantation (Farrell, 2001), "is a tactic norm entrepreneurs employ to institutionalize a new norm by associating it with a preexisting norm in the same issue area, which makes a similar prohibition or injunction" (Acharya, 2004, p. 244), and ensures that a new (global) norm will resonate with a pre-existing (domestic) normative framework (Carpenter, 2007). For example, advocacy efforts to generate a new norm against chemical weapons were helped by associating these weapons with the pre-existing norm against poison (Price, 1997). In addition, the campaign to generate an international norm against antipersonnel (AP) land mines involved efforts to associate AP land mines with delegitimized practices of warfare, including chemical and biological weapons (Price, 1998). A resulting ban on AP land mines was made possible by placing these weapons in the same normative category as other "taboo" weapons. Grafting is thus a tactic that takes time and can be viewed in contrast to "radical transplantation" or "norm displacement" (Farrell, 2001). Similarly, *pruning* involves making adjustments to an outside norm "that make it compatible with local beliefs and practices, without compromising its core attributes" (Acharya, 2004, p. 250). Unlike grafting, however, in which local stakeholders aim to redefine an outside norm by associating it with specific pre-existing norms and practices, pruning consists of selecting those elements of an outside norm that are compatible with the pre-existing local normative framework, and rejecting those elements that are not compatible (Acharya, 2004).

Framing represents another dynamic process in norm translation and localization. Using acts of framing, local norm entrepreneurs borrow an outside (global) norm and present it in a way that emphasizes its value to local stakeholders (Acharya, 2004). Local norm entrepreneurs engage in framing because "the linkages between existing norms and emergent norms are not often obvious and must be actively constructed by proponents of new norms" (Finnemore & Sikkink, 1998, p. 908). The impact of specific frames will depend upon how well they resonate with pre-existing local normative frameworks.

Raymond et al. (2014), in investigating normative change, develop these ideas and extend our understanding of framing processes by introducing the concepts of normative reframing and normative innovation, and connecting them to institutional change. When analyzing institutional change, it is helpful to conceptually distinguish between formal versus informal institutions. On the one hand, informal institutions are "*socially shared rules, usually unwritten, that are created, communicated, and enforced outside of officially sanctioned channels,*" whereas "*formal* institutions are rules and procedures that are created, communicated, and enforced through channels widely accepted as official" (Helmke & Levitsky, 2004, p. 727, emphasis in original). Focusing on intractable problems, Raymond et al. (2014) argue that norms can create pathways for meaningful change in both formal and/or informal institutions due to their powerful influence over human behavior (see also Raymond & Weldon, 2014; Chappell, 2014). Similar to the localization literature, Raymond et al. (2014) examine social and political change in cases where congruence is initially low. They demonstrate that norms can be powerful advocacy tools that mobilize change, and advocates can consciously and intentionally use norm-based strategies when they promote new norms through normative reframing and/or normative innovation that compete with certain pre-existing norms (Raymond et al., 2014).

Both normative reframing and normative innovation begin with *foregrounding* – a process in which advocates highlight and raise awareness about a problematic pre-existing norm. Through foregrounding, advocates aim to challenge a status-quo problematic norm by calling attention to its harmful impact in order to undermine the norm's appeal and prompt others to rethink the implications of the current norm. Advocates accomplish this weakening of an existing norm by making arguments about how the norm is "unreasonable, harmful, inconsistent with other norms, or inappropriately applied" to a particular issue or problem (Raymond et al., 2014, p. 200). Once the norm supporting the status-quo is weakened, advocates can then make use of a norm-based strategy (normative reframing or normative innovation) to replace it with a new norm that supports the institutional change they are promoting (Raymond & Weldon, 2014; Raymond et al., 2014; Raymond, 2016).

Normative reframing is a framing act in which advocates seek to replace a foregrounded norm with a new norm that they deem to better fit the issue at hand, thereby recasting the issue in a new light. Because of the inherent ambiguity of norms, there are often different norms that can be applied to a given problem (Raymond, 2016). Through normative reframing, advocates call attention to the poor fit of the foregrounded norm to the issue at hand, or how it conflicts with other norms applicable to the issue and demonstrate how the newly-proposed alternative norm is a better fit to the issue (Raymond & Weldon, 2014; Raymond et al., 2014; Raymond, 2016). Prior research has emphasized that normative reframing is likely to be effective when it is based upon a strong alternative norm – that is, a norm that is persuasive, broadly and strongly held, can influence behavior, and entails meaningful sanctions when violated (Raymond et al., 2014; Raymond, 2016). Human rights norms are prime examples of norms that are considered strong (Raymond et al., 2014; Raymond, 2016). Normative reframing can be illustrated through the example of violence against women (VAW). In several countries, advocates began by foregrounding the existing gender norms that were facilitating VAW as an acceptable practice in an attempt to weaken those norms (Raymond et al., 2014). Advocates were then able to leverage international human rights and gender equality norms to apply them to the issue of VAW. By reframing VAW as a human rights issue as opposed to a private matter, advocates could pave the way for institutional change in positioning VAW as a matter of public policy (Raymond et al., 2014). In sum, applying alternative norms via normative reframing can redefine the way individuals think about an issue and can suggest the need for new policy prescriptions, which can lead to significant institutional change.

Normative innovation is another framing strategy that entails introducing a new norm. After a pre-existing norm has been foregrounded, advocates can develop and diffuse an entirely new or novel norm in its place (Raymond et al., 2014). In other words, they "create and promote an entirely new norm to promote alternative institutional arrangements, both formal and informal" (Raymond et al., 2014, p. 198). Normative innovation therefore involves the complete rejection of a pre-existing norm that supports the status-quo. When a status-quo norm is evidently applicable to the issue at hand and cannot be foregrounded because it is a poor fit, then advocates may engage in acts of normative innovation to fashion a new norm. Advocates must engage in processes of deliberation in order to bring about and justify a new norm (Raymond et al., 2014). For example, in the case of VAW norm change, activists engaged in deliberation to foreground what they deemed to be undesirable social norms that supported violence against women, and then introduced a new concept, that of VAW, to cover all forms of violence against women including sexual assault, domestic violence, female genital mutilation, and more. This illustrates how normative innovation often entails the creation of new concepts and categories (Raymond et al., 2014). In addition, activists worked to introduce and diffuse new norms regarding masculinity in order to support their goals (Raymond & Weldon, 2014; Raymond et al., 2014). In general, advocates are able to make use of these norm-based acts of framing due to the ambiguity of norms and the potential for a number of different norms to apply to a particular issue or problem. Although these framing strategies are often analyzed in the context of sudden policy change, they can also be more generally applied to better understand how global norms are translated to the local level within a given policy area.

2.3 Issue Framing

As many of the adaptive processes discussed in the previous section involve acts of framing, this section places those dynamics into the broader context of the issue framing literature. In general, the goal of framing is to "select some aspects of a perceived reality and make them more salient in a communicating text, in such a way as to promote a particular problem definition, causal interpretation, moral evaluation, and/or treatment recommendation" (Entman, 1993, p. 52; see also Benford & Snow, 2000). In other words, an issue frame reflects a particular interpretation of an issue (Goffman, 1974) that emphasizes a specific problem definition or causal understanding. When advocates engage in the construction of issue frames, they propose new ideas and meanings – what Benford and Snow (2000, p. 613) call "meaning work" – whose purpose is to influence how individuals process information about the issue. Issue frames accomplish this by directing attention to specific aspects of a problem or issue, thereby highlighting segments of the issue that are in-frame and out-of-frame, and relating a particular narrative to those aspects of an issue that are in-frame. Thus, issue frames are a significant part of processes of interpretation, as they influence how a given issue is perceived and understood (Snow, 2013). In general, advocates rely on issue frames to generate agreement about the nature of an issue and how to most effectively address it, as well as to mobilize action to support the desired solution (Benford & Snow, 2000; Snow, 2013).

According to Matthes (2012, p. 248) the issue framing process can be divided into three phases, with dynamics specific to each: (1) the construction of frames (he focuses on frame construction by political elites, but civil society actors also construct frames), (2) the application of frames, and (3) the effects of frames on individual attitudes and behavior. There has been a great deal of framing research examining framing effects (phase 3), however, the first and second phases of the framing process remain relatively understudied (Ferree et al., 2002; Hänggli, 2012; Helbling, 2014). As a result, our understanding of how stakeholders involved in an issue engage in the construction and application of issue frames remains relatively limited. The cases presented in this book aim to help address this gap by examining how global environmental norms are used to construct issue frames and, furthermore, how those frames are applied to generate shared understandings of local, context-specific environmental issues in France.

Frame resonance, or the success of a particular frame, is influenced by several factors. Klandermans (1984), for example, posits that the success of a social movement campaign can be sufficiently determined by consensus and action mobilization. Snow and Benford (1988), however, argue that scholars must consider the interactive and dynamic relationships that characterize aspects of mobilization and political action. To achieve frame resonance, they argue that three types of framing are necessary: diagnostic framing, prognostic framing, and motivational framing (Snow & Benford, 1988). Diagnostic framing involves two components: identifying the problem/issue, and attributing blame or causality for it (Snow & Benford, 1988). For example, the injustice frame represents one type of diagnostic framing that focuses on identifying a victim and highlighting aspects of victimization (Snow & Benford, 1988). The amplification of victimization that an injustice frame permits has been associated with impactful frame resonance (Anheier et al., 1998; Payne, 2001). Prognostic framing involves identifying solutions to a given problem or issue, while at the same time identifying resources, targets, tactics, and strategies (Snow & Benford, 1988). It is via prognostic framing that we are likely to see differences in approaches arise among actors involved in the issue. Although stakeholders may share a common problem identification, they may not all agree on the same solution to the problem (Benford & Snow, 2000). Similarly, previous research has shown that different stakeholders involved in an issue may draw attention to or away from different norms that apply to the situation, because they do not all share the same social and political context, and background experiences (e.g., Wiener, 2016). Different stakeholders can hold widely divergent perceptions regarding the significance of different norms that are potentially applicable to an issue (Wiener, 2016). To mobilize action around an issue, both diagnostic and prognostic framing are necessary. However, they are not sufficient to garner mass support among publics. For that, motivational framing must be utilized to provide a reason for undertaking action (Snow & Benford, 1988). It is here that we are likely to see the use of specific norms being applied to an issue. For example, social justice norms are often invoked by NGOs active in international development to mobilize resources. In doing so, they may engage in grafting by associating their goals with pre-existing international norms and standards, including United Nations (UN) declarations (Joachim, 2007). Lastly, the framing of an issue is also strategic in that different frames can be used to appeal to different audiences. As Keck and Sikkink (1998) explain, "Land use rights in the Amazon, for example, took on an entirely different character and gained quite different allies viewed in a deforestation frame than they did in either social justice or regional development frames" (p. 17). Frame resonance, in short, relates to the ability to influence broader public understandings. This means that a given frame must be internally coherent and must be a good match to the wider political culture in which its use takes place (Snow & Benford, 1988). Keck and Sikkink (1998, p. 18) relate the different types of framing to "information politics" and processes of persuasion. They summarize that "An effective frame must show that a given state of affairs is neither natural nor accidental, identify the responsible party or parties, and propose credible solutions" (Keck & Sikkink, 1998, p. 19). The messages required to accomplish this must be clear and powerful, and connect to broadly shared principles (Keck & Sikkink, 1998) such as environmental sustainability and preservation in the cases examined in this book.

2.4   Actor Constellations

The literature on norm diffusion identifies specific constellations of actors that engage in the creation and spread of norms. For example, Keck and Sikkink (1998) identify transnational advocacy coalitions (TANs) as important "norm translators" that facilitate and participate in translational norm change. An important feature of TANs is their ability to "mobilize information strategically to help create new issues and categories and to persuade, pressure, and gain leverage over much more powerful organizations and governments" (Keck & Sikkink, 1998, p. 2). Within TANs, NGOs are primary actors, highlighting the strong relevance of non-state actors in processes of norm translation and diffusion. Other actors involved in TANs include components of regional and international organizations. According to Keck and Sikkink's "boomerang pattern" (1998, p. 12), the targets of TANs are norm-violating domestic governments. Through the joint actions of domestic and transnational actors, TANs seek to change state behavior. The concept of transnational advocacy is useful for identifying networks that comprise domestic and transnational NGOs, international experts, and elements of international organizations whose joint efforts lead to norm diffusion.

Less explored in processes of norm diffusion, however, are constellations of domestic-level norm takers, or norm entrepreneurs, who translate and transmit international norms within a particular domestic context. As a result of a strong focus in the literature on the role of national governments as norm transmitters, our understanding of the ways in which sub-national domestic actors diffuse and implement norms locally remains limited (Winanti & Hanif, 2020). Few studies have examined the contextual factors that shape norm-takers' responses to global norms, yet broader theories of political opportunity structures, developed in the social movements literature, point to such factors as government openness, state capacity, and the presence of supportive national or sub-national elites as important structural aspects that enable norm localization. In general, political opportunity structure theories focus on relevant dimensions of the political environment "that provide incentives for people to undertake collective action" (Meyer & Minkoff, 2004, p. 1459) and create opportunities that enable the translation of global norms (Gleditsch & Ruggeri, 2010). These aspects that characterize an open political and institutional environment may relate to the concept of congruence in the norm localization literature by facilitating norm adoption, translation, and/or implementation. In general, efforts by constellations of sub-national actors to promote global-level

environmental norms and mainstream them into their local institutions have not yet
been systematically or explicitly investigated in the norm localization literature,
particularly within a context of high congruence. The case studies that follow in this
book aim to advance our understanding of such local and contextual dynamics.

2.5   Norms and Impact Translation

In her analysis of the repercussions of the Convention on the Elimination of All
Forms of Discrimination Against Women (CEDAW), Zwingel (2012, p. 125) develops the concept of impact translation to describe how constellations of actors use
international norms to wield influence at the domestic level. Drawing on prior
empirical research, she explains how analysis of impact translation requires placing
"emphasis on thoroughly characterizing both the actor constellation and the context
relevant for the action of translation, as this context often determines strategies and
outcomes" (Zwingel, 2012, p. 125). Relatedly, research focusing on the "vernacu-
larization" of international gender norms examines the translation of international
gender norms in the domestic context (Zwingel, 2012). In summarizing the findings
of this research, Zwingel shows how the local context into which the international
norm is translated heavily influences the process and extent of norm translation
(2012, p. 125). For example, research examining global gender equality norms in
the context of China and the United States finds that "some of the translating actors
do not attempt to use international norms for legal or social transformation, but they
adapt the international impulses to the framework already in place" (Zwingel, 2012,
p. 125). This incomplete form of impact translation is referred to as "contextualization" in the Chinese case study (Liu et al., 2009) and as "domestication" in the US
context (Rosen & Yoon, 2009). In other cases, impact translation is more extensive,
for example, in casting a novel international light on long-standing local traditions
in order to reinvent them (Zwingel, 2012). Finally, impact translation can consist of
compartmentalization, a process in which only specific parts or aspects of an international norm that are the most likely to find widespread acceptance are promoted
at the domestic level (Rajaram & Zararia, 2009), similar to what the norm localization literature calls grafting. In the empirical cases in this book, the concept of
impact translation is used as a lens through which I examine the extent of diffusion
of global twilight norms in local settings in France.

2.6   Summary and Conclusion

The theoretical framework articulated in this chapter synthesizes acts of norm localization with theoretical formulations about norm-based change and issue framing to
investigate the dynamic aspects of norm diffusion. Specifically, concepts such as
grafting and pruning from the localization literature are incorporated alongside

normative reframing and normative innovation dynamics from the literature on norm-based institutional change. Additionally, the theoretical framework draws upon the literature on frame resonance to integrate specific types of framing – diagnostic framing, prognostic framing, and motivational framing. In combining these elements, the empirical analyses will be able to shed light on the construction and application of issue frames, which have been less prioritized in research compared to framing effects.

Turning to actors and settings, the identification of constellations of local-level actors who adapt and embed global norms to a specific domestic context contributes to the literature that highlights the relevance of non-state actors in processes of norm translation and diffusion. Relatedly, the concept of impact translation is integrated as a means of theoretically emphasizing the domestic contextual setting into which global norms are translated, and as a lens through which to view the degree of global norm translation in local settings. Central to the framework is the assumption that pure norm adoption, whereby a global norm is applied to the domestic context intact and unchanged, is rare and unlikely. We can more likely expect that global norms will be modified and shaped to the domestic context in processes of norm diffusion and translation, due to the inherent ambiguity of norms and the applicability of multiple norms to a given issue. This ambiguity also opens up space to observe congruence-building acts in translating global norms to the domestic context. Finally, the theoretical framework is relevant in the context of France, where congruence between global-level environmental norms and domestic norms can be regarded as high, because different sub-national stakeholders and actor constellations can potentially interpret a given norm differently depending upon their interests and the context in which they operate, opening up the possibility to observe the dynamic processes of norm translation and localization identified above.

References

Acharya, A. (2004). How ideas spread: Whose norms matter? Norm localization and institutional change in Asian regionalism. *International Organization, 58*(2), 239–275. https://doi.org/10.1017/S0020818304582024

Anheier, H. K., Neidhardt, F., & Vortkamp, W. (1998). Movement cycles and the Nazi party: Activities of the Munich NSDAP, 1925–1930. *American Behavioral Scientist, 41*(9), 1262–1281. https://doi.org/10.1177/0002764298041009006

Benford, R. D., & Snow, D. A. (2000). Framing processes and social movements: An overview and assessment. *Annual Review of Sociology, 26*(1), 611–639. https://doi.org/10.1146/annurev.soc.26.1.611

Boehmke, F. J., & Witmer, R. (2004). Disentangling diffusion: The effects of social learning and economic competition on state policy innovation and expansion. *Political Research Quarterly, 57*(1), 39–51. https://doi.org/10.1177/106591290405700104

Botcheva, L., & Martin, L. L. (2001). Institutional effects on state behavior: Convergence and divergence. *International Studies Quarterly, 45*(1), 1–26. https://doi.org/10.1111/0020-8833.00180

Cardenas, S. (2007). *Conflict and compliance: State responses to international human rights pressure.* University of Pennsylvania Press.

Carpenter, C. R. (2007). Setting the advocacy agenda: Theorizing issue emergence and nonemergence in transnational advocacy networks. *International Studies Quarterly, 51*(1), 99–120. https://doi.org/10.1111/j.1468-2478.2007.00441.x

Chappell, L. (2014). Conflicting institutions and the search for gender justice at the international criminal court. *Political Research Quarterly, 67*(1), 183–196. https://doi.org/10.1177/1065912913507633

Checkel, J. T. (1999). Norms, institutions, and national identity in contemporary Europe. *International Studies Quarterly, 43*(1), 83–114. https://doi.org/10.1111/0020-8833.00112

Checkel, J. T. (2001). Why comply? Social learning and European identity change. *International Organization, 55*(3), 553–588. https://doi.org/10.1162/00208180152507551

Cortell, A. P., & Davis, J. W. (1996). How do international institutions matter? The domestic impact of international rules and norms. *International Studies Quarterly, 40*(4), 451–478. https://doi.org/10.2307/2600887

Cortell, A. P., & Davis, J. W. (2000). Understanding the domestic impact of international norms: A research agenda. *International Studies Review, 2*(1), 65–87. https://doi.org/10.1111/1521-9488.00184

DiMaggio, P. J., & Powell, W. W. (1983). The iron cage revisited: Institutional isomorphism and collective rationality in organizational fields. *American Sociological Review, 48*(2), 147–160. https://doi.org/10.2307/2095101

Eimer, T. R., Lütz, S., & Schüren, V. (2016). Varieties of localization: International norms and the commodification of knowledge in India and Brazil. *Review of International Political Economy, 23*(3), 450–479. https://doi.org/10.1080/09692290.2015.1133442

Elkins, Z., & Simmons, B. A. (2005). On waves, clusters, and diffusion: A conceptual framework. *The Annals of the American Academy of Political and Social Science, 598*(1), 33–51. https://doi.org/10.1177/0002716204272516

Entman, R. M. (1993). Framing: Toward clarification of a fractured paradigm. *Journal of Communication, 43*(4), 51–58. https://doi.org/10.1111/j.1460-2466.1993.tb01304.x

Farrell, T. (2001). Transnational norms and military development: Constructing Ireland's professional army. *European Journal of International Relations, 7*(1), 63–102. https://doi.org/10.1177/1354066101007001003

Ferree, M. M., Gamson, W. A., Gerhards, J., & Rucht, D. (2002). *Shaping abortion discourse. Democracy and the public sphere in Germany and the United States.* Cambridge University Press.

Finnemore, M., & Sikkink, K. (1998). International norm dynamics and political change. *International Organization, 52*(4), 887–917. https://doi.org/10.1162/002081898550789

Gleditsch, K. S., & Ruggeri, A. (2010). Political opportunity structures, democracy, and civil war. *Journal of Peace Research, 47*(3), 299–310. https://doi.org/10.1177/0022343310362293

Goffman, E. (1974). *Frame analysis: An essay on the organization of experience.* Harper & Row.

Grillot, S. R. (2011). Global gun control: Examining the consequences of competing international norms. *Global Governance, 17*(4), 529–555. http://www.jstor.org/stable/23104290

Gustafsson, M. T., & Scurrah, M. (2019). Strengthening subnational institutions for sustainable development in resource-rich states: Decentralized land-use planning in Peru. *World Development, 119*, 133–144. https://doi.org/10.1016/j.worlddev.2019.03.002

Hafner-Burton, E. M. (2005). Trading human rights: How preferential trade agreements influence government repression. *International Organization, 59*(3), 593–629. https://doi.org/10.1017/S0020818305050216

Hänggli, R. (2012). Key factors in frame building: How strategic political actors shape news media coverage. *American Behavioral Scientist, 56*(3), 300–317. https://doi.org/10.1177/0002764211426327

Helbling, M. (2014). Framing immigration in western Europe. *Journal of Ethnic and Migration Studies, 40*(1), 21–41. https://doi.org/10.1080/1369183X.2013.830888

Helmke, G., & Levitsky, S. (2004). Informal institutions and comparative politics: A research agenda. *Perspectives on Politics, 2*(4), 725–740. https://doi.org/10.1017/S1537592704040472

Joachim, J. M. (2007). NGOs and UN agenda setting political opportunities, mobilizing structures, and framing strategies. In J. M. Joachim (Ed.), *Agenda setting, the UN, and NGOs* (pp. 15–40). Georgetown University Press.

Keck, M. E., & Sikkink, K. (1998). *Activists beyond borders*. Cornell University Press.

Kersbergen, K. V., & Verbeek, B. (2007). The politics of international norms: Subsidiarity and the imperfect competence regime of the European union. *European Journal of International Relations, 13*(2), 217–238. https://doi.org/10.1177/1354066107076955

Klandermans, B. (1984). New social movements and resource mobilization: The European and the American approach. *International Journal of Mass Emergencies and Disasters, 4*(2), 13–37. http://www.ijmed.org/articles/466/download/

Legro, J. W. (1997). Which norms matter? Revisiting the 'failure' of internationalism. *International Organization, 51*(1), 31–63. https://doi.org/10.1162/002081897550294

Liu, M., Hu, Y., & Liao, M. (2009). Travelling theory in China: Contextualization, compromise, and combination. *Global Networks, 9*(4), 529–554. https://doi.org/10.1111/j.1471-0374.2009.00267.x

Matthes, J. (2012). Framing politics: An integrative approach. *American Behavioral Scientist, 56*(3), 247–259. https://doi.org/10.1177/0002764211426324

Meyer, D. S., & Minkoff, D. C. (2004). Conceptualizing political opportunity. *Social Forces, 82*(4), 1457–1492. https://doi.org/10.1353/sof.2004.0082

Payne, R. A. (2001). Persuasion, frames and norm construction. *European Journal of International Relations, 7*(1), 37–61. https://doi.org/10.1177/1354066101007001002

Price, R. (1997). *The chemical weapons taboo*. Cornell University Press.

Price, R. (1998). Reversing the gun sights: Transnational civil society targets land mines. *International Organization, 52*(3), 613–644. https://doi.org/10.1162/002081898550671

Rajaram, N., & Zararia, V. (2009). Translating women's human rights in a globalizing world: The spiral process in reducing gender injustice in Baroda, India. *Global Networks, 9*(4), 462–484. https://doi.org/10.1111/j.1471-0374.2009.00264.x

Raymond, L. (2016). *Reclaiming the atmospheric commons: The regional greenhouse gas initiative and a new model of emissions trading*. MIT Press.

Raymond, L., & Weldon, S. L. (2014). Mini-symposium introduction: Informal institutions and 'intractable' global problems. *Political Research Quarterly, 67*(1), 181–182. https://doi.org/10.1177/1065912913518662

Raymond, L., Weldon, S. L., Kelly, D., Arriaga, X. B., & Clark, A. M. (2014). Making change: Norm-based strategies for institutional change to address intractable problems. *Political Research Quarterly, 67*(1), 197–211. https://doi.org/10.1177/1065912913510786

Risse-Kappen, T. (1994). Ideas do not flow freely: Transnational coalitions, domestic structures, and the end of the cold war. *International Organization, 48*(2), 185–214. http://www.jstor.org/stable/2706930

Rosen, M. S., & Yoon, D. H. (2009). "Bringing coals to Newcastle"? Human rights, civil rights, and social movements in New York City. *Global Networks, 9*(4), 507–528. https://doi.org/10.1111/j.1471-0374.2009.00266.x

Sandholtz, W. (2008). Dynamics of international norm change: Rules against wartime plunder. *European Journal of International Relations, 14*(1), 101–131. https://doi.org/10.1177/1354066107087766

Savery, L. (2007). *Engendering the state: The international diffusion of women's human rights*. Routledge.

Snow, D. A. (2013). Framing and social movements. In D. A. Snow, D. della Porta, B. Klandermans, & D. McAdam (Eds.), *The Wiley-Blackwell encyclopedia of social and political movements* (pp. 470–474). Malden, MA.

Snow, D. A., & Benford, R. (1988). Ideology, frame resonance and participant mobilization. In B. Klandermans, H. Kriesi, & S. Tarrow (Eds.), *From structure to action: Comparing social movement research across cultures* (pp. 197–217). Jai Press.

Sunstein, C. R. (1996). Social norms and social roles. *Columbia Law Review, 96*(4), 903–968. https://doi.org/10.2307/1123430

Welsh, J. M. (2013). Norm contestation and the responsibility to protect. *Global Responsibility to Protect, 5*(4), 365–396. https://doi.org/10.1163/1875984X-00504002

Wiener, A. (2009). Enacting meaning-in-use: Qualitative research on norms and international relations. *Review of International Studies, 35*(1), 175–193. https://doi.org/10.1017/S0260210509008377

Wiener, A. (2016). Contested norms in inter-national encounters: The 'turbot war' as a prelude to fairer fisheries governance. *Politics and Governance, 4*(3), 20–36. https://doi.org/10.17645/pag.v4i3.564

Winanti, P. S., & Hanif, H. (2020). When global norms meet local politics: Localising transparency in extractive industries governance. *Environmental Policy and Governance, 30*(5), 263–275. https://doi.org/10.1002/eet.1907

Winston, C. (2018). Norm structure, diffusion, and evolution: A conceptual approach. *European Journal of International Relations, 24*(3), 638–661. https://doi.org/10.1177/1354066117720794

Young, O. R. (1999). Regime effectiveness: Taking stock. In O. R. Young (Ed.), *The effectiveness of international environmental regimes: Causal connections and behavioral mechanisms* (pp. 249–280). MIT Press.

Zwingel, S. (2012). How do norms travel? Theorizing international women's rights in transnational perspective. *International Studies Quarterly, 56*(1), 115–129. https://doi.org/10.1111/j.1468-2478.2011.00701.x

Chapter 3
The Cerbère-Banyuls Marine Nature Reserve

This chapter examines the case of the Cerbère-Banyuls Marine Nature Reserve (MNR). It identifies specific twilight norms and begins to trace some of the ways that these global-level environmental norms have been used to help justify the creation of the MNR, to situate its management and regulation into a wider context, and to help substantiate the importance of the current goals and objectives associated with the ongoing development and maintenance of the MNR. The examples presented in this chapter are not intended to be exhaustive or to represent a full account of the discourse surrounding this MNR. Rather, the objectives are to provide a starting point in showcasing specific examples that illustrate how certain global environmental norms have been (and are being) used and adapted to the local context, and to highlight examples of framing, norm linkages, and contextualization with the aim of placing some of the discourse specific to this MNR into a broader theoretical context.

The Cerbère-Banyuls MNR was founded in 1974, making it the oldest marine reserve in France. Located in southern France on the Catalan coast, the marine reserve is situated in the western part of the Gulf of Lion, bordering the rocky coast of the department of Pyrénées-Orientales. It is located 35 km south of Perpignan and 2 km north of France's border with Spain (Le Département des Pyrénées-Orientales, n.d.-a). This MNR represents a relatively small area covering 650 ha of sea, and stretches over 6.5 km of coastline between Banyuls-sur-Mer and Cerbère, and up to 1.5 miles out to sea (Le Département des Pyrénées-Orientales, n.d.-a). It is composed of several distinct parts, including (1) the partial protection zone, where human activities such as fishing are regulated, while others including spearfishing are prohibited, (2) the reinforced protection zone, where all sampling, immersion, and anchoring is prohibited, except when approved for scientific studies by the Advisory Committee of the Reserve, (3) two designated areas for anchorage at Cap l'Abeille and Anse de Peyrefite, and (4) a 250 meter-long underwater trail on Peyrefite beach that can be visited along a marked route, containing five observation stations that represent five different ecosystems: pebbles, Posidonia seagrass,

M. Schnyder, *Global Norms in Local Contexts*, SpringerBriefs in Political Science, https://doi.org/10.1007/978-3-031-41108-3_3

blocks, faults, and drop-offs (Le Département des Pyrénées-Orientales, n.d.-b, n.p.; Le Département des Pyrénées-Orientales, 2020a). These distinct parts of the MNR correspond to two broad levels of protection: one where human activities are regulated (consisting of approximately 600 ha) and another where all human activities are prohibited (consisting of 65 ha) (Le Département des Pyrénées-Orientales, 2020a). In general terms, an MNR guarantees the protection and diversity not only of the animal and plant species specific to it, but also of the natural environment in which those species live. To this end, the objectives of the Cerbère-Banyuls MNR include conservation of specific habitats and diversity, controlling human activities in order to make site visits compatible with the conservation objectives, and promoting public awareness through the promotion of education and cultural interest (Le Département des Pyrénées-Orientales, n.d.-b, n.p.).

In 2014, the Cerbères-Banyuls MNR was one of the first marine protected areas to be recognized on the International Union for Conservation of Nature (IUCN) Green List of Protected and Conserved Areas, and in 2018 it was reaffirmed on the Green List. More than 40 years since its formal creation, the local flora and fauna have recovered, and marine and coastal species continue to thrive. In fact, "over 1200 animal species and around 500 plant species have been described in the Natural Reserve," among which "49 are registered as protected species under a national, European or international treaty" (IUCN, 2022, n.p.). Since 1977, the MNR has been managed with the aim of reconciling the protection of the seabed, the preservation of underwater species, and sustainable management of socio-economic activities (Payrot et al., 2014a, p. 14). Arriving at this success point, however, necessitated the coordination and involvement of multiple stakeholders, as the following section describes.

3.1 Background and Actor Constellations

Acting as norm entrepreneurs[1] in pushing to change the existing standards of behavior, momentum to create a marine reserve initially began due to concern expressed by the local community over excessive "pollution and degradation from tourism and fishing" in the area (Fleming, 2016, n.p.). This initial environmental degradation had several contributing factors. Starting in the mid-twentieth century, several dynamics began to have a negative impact on the local ecological balance. For one, the development of tourism and recreation led to an increase in fishing, and more boating and scuba diving, which, over time, resulted in the depopulation of seabeds (Le Département des Pyrénées-Orientales, n.d.-a, n.p; Payrot et al., 2014a, p. 12). In addition, flora and fauna were being weakened and destroyed by an increase in pollution being discharged directly into the sea, including waste oil from marine

---

[1] Norm entrepreneurs refer to those individuals interested in changing existing norms (Sunstein, 1996).

engines, and the modernization of the fishing industry in the 1960s also took its toll (Le Département des Pyrénées-Orientales, n.d.-a, n.p.).

A range of stakeholders were involved in the initial creation and subsequent management of the marine reserve. In the mid-1960s, the municipality of Cerbère initiated the MNR project, which the neighboring municipality of Banyuls-sur-Mer also joined (Le Département des Pyrénées-Orientales, n.d.-a, n.p.). In 1971, the Observatoire océanologique de Banyuls-sur-Mer, also known as the Arago Laboratory (a marine station located in Banyuls-sur-Mer), became involved in the effort by presenting a scientific report that underscored the need for the creation of a marine biological reserve to protect certain species that were becoming particularly endangered (Payrot et al., 2014a, p. 12). Following this, the interministerial decree of 26 February 1974 signed by the Minister of Transport and the Minister of the Environment formally created France's first MNR (Le Département des Pyrénées-Orientales, n.d.-a, n.p.). Then in 1978, with the consent of the Banyuls Fishing Prud'homie (fishing collective), the Maritime Prefect for the Mediterranean issued a decree to establish the "scientific cantonment of Cap Rédéris" with the goal of strengthening the status of the MNR by creating a reinforced protection zone (Le Département des Pyrénées-Orientales, n.d.-a, n.p.). Finally, as a result of wide-ranging consultations with users of the marine reserve, scientists, and the relevant national and local administrations, Decree 90–790 came into force in September of 1990, replacing the 1974 decree (Le Département des Pyrénées-Orientales, n.d.-a, n.p.). This decree marked an important point in the life of the MNR, as it established new governance in the management of the site through greater involvement of the users of the marine reserve and, further, sustained the reinforced protection zone of Cape Rédéris (Le Département des Pyrénées-Orientales, n.d.-a, n.p.).

Upon its creation on February 26, 1974, the interministerial decree provided for the establishment of a management council comprising local actors, including fishermen and members of local administration bodies, and was chaired by the Director of the Arago Laboratory (Payrot et al., 2014a, p. 16). In 1977, the General Council of the Pyrénées-Orientales[2] (the regional government) agreed to take over the management of the MNR and has since provided resources for monitoring the site (Payrot et al., 2014a, p. 17), with the aim of protecting the seabed and marine species, and sustainably managing human activities and development. Today, the MNR also has an Advisory Committee chaired by the Prefect of the Department Pyrénées-Orientales, which takes part in its governance and management. Members of the Advisory Committee, appointed for a term of 3 years, include local authorities, users of the MNR, representatives of public administrations, representatives of associations for the protection of nature, and qualified scientific personnel (Payrot et al., 2014a, p. 17). Together with the Departmental Council, the Advisory Committee is a key norm-making body in the sense that it has a voice in shaping and developing the norms that apply to the MNR. In addition, in 2000 a Scientific Council was established to bring independent scientific expertise to the

---

[2] Now called the Departmental Council of the Pyrénées-Orientales.

management of the MNR. Today, the Scientific Council, which can also be regarded as a norm-making body, consists of 14 members and meets 3–4 times a year (Payrot et al., 2014a, p. 18). Overall, these committees represent a diverse range of stakeholders and provide the framework for multi-stakeholder involvement in the governance and management of the MNR. Both the initial push to create the MNR, as well as its ongoing governance and management, have helped to create a local discourse that reflects specific environmental norms and principles that have been adapted to the local preservation-focused setting.

3.2   Environmental Norms

This section traces the global environmental twilight norms that present themselves in the context of the Cerbère-Banyuls MNR. In identifying these global norms, it also assesses some of the ways they have been adapted to fit within a particular local context, providing an interpretation and discussion of how certain global norms are used in the setting of this MNR to frame environmental conservation, sustainability, and biodiversity preservation.

Three global twilight norms become evident in examining the case of the Cerbère-Banyuls MNR: sustainable development, intergenerational equity, and common heritage. First, the norm of sustainable development stipulates that the environment must be preserved for future generations while meeting the economic and growth needs of the present (Barral, 2012), and is reflected in writings about the MNR that speak about the management approach, regulation, and coexistence of various types of activities that balance its preservation with its ongoing use. As the regional government details,

After more than 45 years of existence, the Cerbère-Banyuls Marine Reserve has gained recognition both nationally and internationally. Distinguished by many titles (IUCN international green list, Glores award, Specially Protected Area of Mediterranean Importance), it is an example in terms of efficiency and benefits generated in favor of the protection of marine ecosystems and coasts as well as the sustainable development of activities. Its status as a protected marine area on a human scale has indeed made it possible to validate an environmental management methodology. The efforts made over the years have made it possible to make this site an open-air laboratory for scientists, but also a space where different activities coexist while respecting the environment. (Le Département des Pyrénées-Orientales, n.d.-c, n.p.)

In general, the stakeholders involved in the MNR's management and monitoring frequently underscore the above-mentioned idea of "coexistence," using it as a way of framing what sustainable development means in this particular local context. The sustainable development norm helps to explain how human activities such as "fishing, yachting and diving activities are regulated to coexist with respect for the environment," as explained by the regional government that manages the reserve (Le Département des Pyrénées-Orientales, 2020b, p. 2). A Cerbère-Banyuls MNR brochure reinforces this idea of balancing practical use with preservation: "Everything is done to keep the Reserve a…site…in which different activities coexist with respect for the environment. Monitoring professional and recreational fishing, partnerships with diving structures … improvements of diving sites with the installation of anchorages to limit damage induced by anchors on the seabed are also carried out" (Le Département des Pyrénées-Orientales, 2020a, p. 6). However, the sustainable development norm is perhaps best reflected in documents concerning professional fishing in the MNR. The 2015–2019 Management Plan (Section B) devotes a section to professional fishing, under the heading of maintain[ing] a sustainable practice of nautical activities in the RNMCB.[3] Spearheaded by the Departmental Council,[4] which acts as a norm-maker in this case, and validated by the Scientific Council, it states that:

> In recent years, professional fishing has experienced many developments in the Mediterranean but also in the RNMCB. After being reduced, the number of fishermen authorized to fish in the reserve has seen an increase since 2010. Since 2011, the 15 authorizations issued are all assigned. The regulation of professional fishing in force in the RNMCB (compulsory authorizations, limitation of the authorizations … etc.) makes it possible to reduce levies and ensure that their practice is in line with the conservation of fishery resources of the RNMCB. (Payrot et al., 2014b, p. 89).

Furthermore, the section of the management plan addressing recreational fishing elaborates on the fact that this activity has been increasing in the MNR, as well as on the perimeter of the reserve. As it explains, "Currently more than 1,400 authorizations are issued each year by the administrator. Current recreational fishing regulations in the RNMCB (authorizations required but not limited to date, minimum catch sizes, quotas for certain species, etc.) make it possible to ensure that their practice is in line with the conservation of fishery resources in the RNMC" (Payrot et al., 2014b, p. 89). In general, coexistence framing, reflected in stakeholder discourse involving the need to balance the practical (often economic) usage of this MNR with its preservation, helps to translate and contextualize the broader idea of sustainable development to this particular local setting.

In addition to sustainable development, the norm of intergenerational equity – that the environment and natural resources should be used in a way that ensures their preservation for the benefit of future generations – is reflected in discourse specific to the MNR. This norm is perhaps most strongly and directly evident in sources that recount the MNR's founding and history. For example, one article showcases the relevance of this norm to the local community, which includes key norm entrepreneurs involved in the MNR's establishment:

It was founded in 1974 as the first Natural Marine Reserve in France from local concern for the area because of the pollution and degradation from tourism and fishing. Cerbère is a commune that relies on tourism, so they wanted to get the local

---

[3] The RNMCB is the acronym for the name of the reserve in French: Reserve Naturelle Marine Cerbère-Banyuls.

[4] Formerly the General Council of the Pyrénées-Orientales (currently the Departmental Council of the Pyrénées-Orientales).

community involved and continue to let people to come enjoy its beauty, without harming the environment. The local community wanted people to enjoy the reserve but also keep it intact so that it can be enjoyed in the future. They agreed on basic management guidelines to maintain its natural values over time. (Fleming, 2016, n.p.)

In addition, the same source touches upon the intergenerational equity norm in showcasing the MNR's educational program: "the Reserve provides an outdoors, underwater classroom to educate the future generations. More than 2,500 students visit Cerbères-Banyuls each year to participate in activities and learn how to become environmentally responsible and promote and value the area's preservation for many years to come" (Fleming, 2016, n.p.).

Although other documentation stresses the norm of equity in accessing and utilizing the MNR, the intergenerational aspect is not always explicitly addressed. To briefly illustrate, an information sheet on the Cerbère-Banyuls MNR published by the Department Pyrénées-Orientales speaks of the reserve in the context of "The Mediterranean for All," stating that "The Marine Nature Reserve, far from being a sealed space, is open and accessible to everyone. It welcomes, protects and transmits its wealth to all its users" (Le Département des Pyrénées-Orientales, 2020a, p. 2). Although this does not explicitly invoke the idea of intergenerational equity, it does promote the general equity principle in framing the MNR as a resource accessible to all. This example is consistent with what the impact translation literature refers to as compartmentalization or contextualization – when only certain parts of a norm are promoted in a specific context (Rajaram & Zararia, 2009; Liu et al., 2009). In the case of the Cerbère-Banyuls MNR, the equity principle is more generally translated to various users of the resource across space, as opposed to the intergenerational aspect which invokes equity across time.

The common heritage principle is the third environmental norm evident in discourse surrounding this MNR, representing the idea that certain global commons or elements regarded as beneficial to humanity as a whole should be held in trust for the benefit of humankind as a whole, as opposed to being unilaterally exploited (Egede, 2014). The common heritage notion is reflected by key norm-makers in considering the reasons for promoting environmental sustainability within the MNR. For example, as the Departmental Council explains in discussing the role of the nature reserve,

A Nature Reserve guarantees the protection and diversity of animal and plant species, but also the natural environment in which they live. However, the protection of the environment does not mean the closure of the latter. This is why the human presence is regulated by reconciling as best as possible protection and use of the area. The aim is to limit withdrawals and disturbance of species at the heart of the Reserve and to promote the restoration of the environment, in order to safeguard our natural heritage. (Le Département des Pyrénées-Orientales, 2020a, p. 3).

The concept of "natural heritage" (patrimoine naturel) is useful to consider here in that it helps illustrate how some stakeholders connect and adapt the global norm of common heritage to the local context. Although the term "patrimoine naturel" first appeared in France in the 1960s, the "revolutionary" law of 10 July 1976 really brought it to the fore (Lefeuvre et al., 1981), in stipulating "that it is everyone's duty to safeguard the natural heritage in which they live" (Lefeuvre, 1990, p. 29). The evolution of the notion of patrimoine naturel involved grafting the norm onto international environmental values, and in this sense was closely related to developments at the international level, including the development of the common heritage of humanity notion evident in UNESCO's World Heritage Convention as well as the Convention of the Law of the Sea (Lefeuvre, 1990, p. 41). Specific to the Cerbère-Banyuls MNR, the Departmental Council and the Scientific Council raise the point, in the 2015–2019 Management Plan (Section B), Challenge VIII, of "Ensur[ing] that the practice of human activities in the RNMCB is sustainable, and compatible with conservation objectives of the natural heritage" (Payrot et al., 2014b, p. 86). It further elaborates that "One of the objectives of the RNMCB is to raise awareness and educate as many people as possible in favor of the conservation of the natural heritage it protects, to promote actions that respect the environment but also to discover its riches. These notions, once transmitted and applied, allow everyone to appreciate and respect them all the better" (Payrot et al., 2014b, p. 86). Furthermore, in communications about extending the boundaries of the current MNR, Hermeline Malherbe, President of the Pyrénées-Orientales Department, has emphasized that "We are all responsible for this heritage and required to take a look at its future" (Le Département des Pyrénées-Orientales, n.d.-c, n.p.). Moreover, as promoted and validated by the Scientific Council, the carrying out of scientific studies is sometimes positioned in the context of preserving the natural heritage that the MNR provides: "Numerous scientific studies are being carried out on the perimeter of the RNMCB in order to assess and measure the state of conservation of the natural heritage. These studies constitute a fundamental field of activity in the conservation of biodiversity and the management of the RNMCB" (Le Département des Pyrénées-Orientales, n.d.-d, n.p.). In sum, discourses that position the Cerbère-Banyuls MNR as an important part of the natural heritage of France and articulate a desire to maintain this natural heritage for future generations help norm-makers link the common heritage norm to the local context relevant to the preservation of the MNR.

3.3   Framing

Issue framing refers to "the process by which people develop a particular conceptualization of an issue or reorient their thinking about an issue" (Chong & Druckman, 2007, p. 104). When an issue frame is used, it reflects a specific interpretation of an issue (Goffman, 1974) that emphasizes a particular way of understanding the problem or its causes (Entman, 1993), underscoring "verbal or visual messages that highlight specific dimensions of policy issues to influence individuals' perceptions of those issues" (Andrews et al., 2017, p. 262). As issue frames place attention on particular aspects of the problem or issue in question, they play a key role in processes of interpretation and serve to influence the way that a particular issue is perceived and understood (Snow, 2013a, pp. 470–471).

Recall from the previous chapter that diagnostic framing involves the two elements of identifying the problem/issue, and attributing blame or causality for it (Snow & Benford, 1988), while prognostic framing highlights solutions to the problem or issue while simultaneously identifying resources, targets, tactics, and/or strategies (Snow & Benford, 1988). There are many examples of diagnostic and prognostic framing that can be found in discourse about the Cerbère-Banyuls MNR, including its history, management, and regulations, which focus attention upon ecological problems and recount causal mechanisms that attribute blame. For example, *harm to the natural environment* is the identified problem that motivated the creation of the MNR and the active involvement of the local community, while blame for it is placed on *tourism and fishing*, which cause harm as a result of *increased pollution* and *damage to the natural environment due to boat anchors* (Fleming, 2016; Le Département des Pyrénées-Orientales, n.d.-c). In terms of prognostic framing, the creation of the MNR itself is framed as the solution: "The reserve is a safe zone for many endangered species which are threatened in other areas from pollution (sewage, metals and litter that make their way into the water), fishing (anchors) and impacts from too much human activity and pressure" (Fleming, 2016, n.p.). More precisely, its specific management principles that regulate access to and activities in the area are framed as solutions. For instance, fishermen are required to obtain and carry a license, and they are only permitted to catch certain types of fish (Préfecture Maritime Méditerranée, 2020). Furthermore, anchors and trawling are banned to prevent damage to the environment and to particular species (Préfecture Maritime Méditerranée, 2020). In addition, although scuba diving is permitted, divers must sign an agreement not to touch or leave anything in the protected area, and there are specific quotas set to limit the total number of divers (Préfecture Maritime Méditerranée, 2020). Scientific research in the marine reserve is permitted but must be approved in advance (Préfecture Maritime Méditerranée, 2020). These rules, which were formulated with the input of various stakeholders, including local residents, reflect prognostic framing as they are framed as positive solutions that help "to ensure the conservation or limitation of animal or plant species in the reserve" (La République Française, 1990, p. 3) and thus prevent the disturbance of ecologically sensitive areas or species (Fleming, 2016).

Diagnostic and prognostic framing are often used together to develop a shared understanding of the value that the Cerbère-Banyuls MNR provides when it comes to safeguarding biodiversity and the natural environment, or in other words, a discursive field (Snow, 2013b). When viewed in the context of the concept of *patrimoine naturel*, what emerges is a discursive field that constructs the MNR as an indispensable part of the environmentally-oriented cultural identity of France. The concept of *patrimoine naturel* also serves to situate the MNR within the broader national context and thereby augment its value on a national scale.

Beyond diagnostic and prognostic framing, motivational framing is often used to provide a reason for undertaking action (Snow & Benford, 1988). In the case of the planned extension of the perimeter of the MNR, motivation for action stems from the following rationale: "in the current context of ecological crisis and citizen awareness, the time has come for action and responsibility" (Le Département des

Pyrénées-Orientales, n.d.-c, n.p.). Furthermore, Hermeline Malherbe has declared that "there is an urgent need to halt the loss of terrestrial and marine biodiversity deemed irreversible if nothing is done by 2030" (Le Département des Pyrénées-Orientales, n.d.-c, n.p.). In the local context of the MNR, this type of motivational framing often goes hand-in-hand with the creation of norms that prohibit touch. As touch is a cultural practice (Jewitt et al., 2020, pp. 57–72), new norms regarding the appropriateness of touch are created by the Advisory Committee in the context of the marine reserve: "Avoid touching and overcrowding as it hurts the ecosystem" (Le Département des Pyrénées-Orientales, 2020a, p. 1); "On the underwater trail, you have to swim with your eyes. Do not turn the stones over, do not feed the fish" (Le Département des Pyrénées-Orientales, 2020a, p. 1). Linking norms of touch to norms of ecological preservation is thus an important strategy that the Advisory Committee uses to help manage and regulate human behavior within the MNR. In doing so, stakeholders also engage in grafting by associating their goals and regulations with pre-existing international norms. For example, rules and regulations (including norms of touch) are grafted to the norm of intergenerational equity: "Since 1977, the Department, as a manager, puts every effort to preserve and enhance this natural fragile space so that it transmits its richness to all its users" (Le Département des Pyrénées-Orientales, 2020b, p. 2). In this regard, the framing of the problem and solution, and motivations for action, can be viewed as strategic in that different types of frames, and different norms, are used to help regulate desired behavior and prevent unwanted actions.

3.4 Conclusion

The central goals of this chapter were to identify the extent to which specific global-level environmental twilight norms are evident in the case of the Cerbère-Banyuls Marine Nature Reserve, and to begin to trace how these norms are adapted to this particular local context. In conducting this research, different examples from a range of sources specific to the MNR were assessed. The point of departure was the assumption that local-level discourse surrounding the creation and importance of the MNR, as well as its management, regulation principles, and ongoing development, does not exist in isolation from the broader global-level environmental norms that have informed international environmental jurisprudence. The premise was that by considering the ways in which these global norms are adapted to the local context, we can develop richer and more nuanced accounts of the significance of environmental twilight norms and better understand how they are leveraged to frame, promote, and generally develop a local discourse that centers on environmental conservation and sustainability.

This chapter highlights several key findings. First, there are certain global-level environmental norms that are indeed relevant in this local context. The twilight norms of sustainable development, intergenerational equity, and common heritage are each reflected in the case of the Cerbère-Banyuls MNR and are actively promoted by key norm-making bodies involved in its governance and management. These norms not only help inform the ongoing management approach and approach to regulating human activities in the MNR in a way that balances its practical usage with its preservation, but are also evident in narratives surrounding its creation and history.

In addition, this chapter illustrates how stakeholders have several means of translating these global norms to the local context of the MNR, mainly through the use of framing. Coexistence framing, for example, is often used to emphasize the aim of balancing the practical and economic usage of the MNR with its preservation, which helps to translate and contextualize the broader idea of sustainable development. Further, those involved in the governance and ongoing management of the MNR use the concept of natural heritage (patrimoine naturel) as a heuristic tool to adapt the norms of sustainable development and common heritage, respectively, to the local setting. In this vein, the chapter brought to light discourses that position the Cerbère-Banyuls MNR as holding significant value as part of the natural heritage of France, while also communicating the need to maintain this natural heritage for future generations. Moreover, the norm of intergenerational equity is compartmentalized and contextualized (Rajaram & Zararia, 2009; Liu et al., 2009) to highlight general principles of equity across space more so than across time, promoting the general equity principle in framing the MNR as a resource accessible to all. Finally, stakeholders utilize different types of framing, including diagnostic and prognostic framing, as well as motivational framing, to develop a discourse and shared understanding of the importance and value of the MNR, and to motivate action to help regulate human behavior within it. Through these processes and mechanisms, global environmental norms contribute to the development of a discursive field (Snow, 2013b) that promotes and reinforces protection, conservation, and sustainability in the context of the Cerbère-Banyuls MNR, which is seen by these stakeholders as an indispensable part of the cultural identity of France.

When local stakeholders construct issue frames, they engage in "meaning work" (Benford & Snow, 2000, p. 613) in order to influence how individuals process information about the MNR and to propose new ideas and meanings. This type of meaning construction has been more broadly characterized as "the struggle over the production of mobilizing and countermobilizing ideas and meanings" (Benford & Snow, 2000, p. 613). From this perspective, the local actors involved with the MNR cannot be viewed as simply transmitters of existing normative ideas and meanings that stem from pre-existing (global) structural arrangements or ideologies. Rather, consistent with the localization perspective (Acharya, 2004), local actors are "signifying agents actively engaged in the production and maintenance of meaning for constituents, antagonists, and bystanders or observers" (Benford & Snow, 2000, p. 613, citations omitted). The phenomenon of frame construction is both active and processual, underscoring the agency of MNR stakeholders as it pertains to reality construction. In this sense, this chapter ultimately lends support to localization perspectives that underscore the dynamic aspects involved in congruence-building processes in the domestic arena, as opposed to the view that congruence merely reflects a static fit between norms.

References

Acharya, A. (2004). How ideas spread: Whose norms matter? Norm localization and institutional change in Asian regionalism. *International Organization, 58*(2), 239–275. https://doi.org/10.1017/S0020818304582024

Andrews, A. C., Clawson, R. A., Gramig, B. M., & Raymond, L. (2017). Finding the right value: Framing effects on domain experts. *Political Psychology, 38*(2), 261–278. https://doi.org/10.1111/pops.12339

Barral, V. (2012). Sustainable development in international law: Nature and operation of an evolutive legal norm. *European Journal of International Law, 23*(2), 377–400. https://doi.org/10.1093/ejil/chs016

Benford, R. D., & Snow, D. A. (2000). Framing processes and social movements: An overview and assessment. *Annual Review of Sociology, 26*(1), 611–639. https://doi.org/10.1146/annurev.soc.26.1.611

Chong, D., & Druckman, J. N. (2007). Framing public opinion in competitive democracies. *American Political Science Review, 101*(4), 637–655. https://doi.org/10.1017/S0003055407070554

Egede, E. (2014). *Common heritage of mankind.* Oxford bibliographies online in international law. Retrieved from https://www.oxfordbibliographies.com/view/document/obo-9780199796953/obo-9780199796953-0109.xml

Entman, R. M. (1993). Framing: Toward clarification of a fractured paradigm. *Journal of Communication, 43*(4), 51–58. https://doi.org/10.1111/j.1460-2466.1993.tb01304.x

Fleming, A. (2016, 28 June). *The tourist-proof region of Cerbère-Banyuls.* IUCN (International Union for Conservation of nature) news. Retrieved from https://www.iucn.org/news/protected-areas/201606/tourist-proof-region-cerb%C3%A8re-banyuls

Goffman, E. (1974). *Frame analysis: An essay on the organization of experience.* Harper & Row.

IUCN (International Union for Conservation of Nature). (2022). *Marine Natural Reserve of Cerbère-Banyuls.* Retrieved from https://iucngreenlist.org/sites/marine-natural-reserve-of-cerbere-banyuls%E2%80%AF/

Jewitt, C., Price, S., Leder Mackley, K., Yiannoutsou, N., & Atkinson, D. (2020). *Interdisciplinary insights for digital touch communication* (Human–computer interaction series). Springer. https://doi.org/10.1007/978-3-030-24564-1_4

La République Française. (1990). Décret N°90–790 du 6 septembre 1990 portant création de la réserve naturelle marine de Cerbère-Banyuls (Pyrénées-Orientales). Retrieved from https://www.legifrance.gouv.fr/jorf/id/JORFTEXT000000168716

Le Département des Pyrénées-Orientales. (2020a). *Brochure de présentation de la Réserve Naturelle Marine de Cerbère Banyuls (PDF).* Retrieved from https://www.ledepartement66.fr/quest-ce-que-la-reserve-marine/

Le Département des Pyrénées-Orientales. (2020b). *Cerbère-Banyuls réserve naturelle marine.* Retrieved from https://www.ledepartement66.fr/wp-content/uploads/2020/09/Reservemarine-depliant-2020-2.pdf

Le Département des Pyrénées-Orientales. (n.d.-a). *Découvrir l'histoire de la Réserve Naturelle Marine de Cerbère-Banyuls.* Retrieved from https://www.ledepartement66.fr/decouvrir-lhistoire-de-la-reserve-naturelle-marine-de-cerbere-banyuls/

Le Département des Pyrénées-Orientales. (n.d.-b). *Qu'est-ce que la réserve marine ?* Retrieved from https://www.ledepartement66.fr/quest-ce-que-la-reserve-marine/

Le Département des Pyrénées-Orientales. (n.d.-c). *Extension de la réserve marine de Cerbère-Banyuls.* Retrieved from https://www.ledepartement66.fr/actualite/extension-de-la-reserve-marine-de-cerbere-banyuls/

Le Département des Pyrénées-Orientales. (n.d.-d). *En savoir plus sur les études scientifiques.* Retrieved from https://www.ledepartement66.fr/en-savoir-plus-sur-les-etudes-scientifiques/

Lefeuvre, J. C. (1990). De la protection de la nature à la gestion du patrimoine naturel. In H. P. Jeudy (Ed.), *Patrimoines en folie, new edition [online]* (pp. 29–75). Éditions de la Maison des sciences de l'homme. https://doi.org/10.4000/books.editionsmsh.3778

Lefeuvre, J. C., Raffin, J. P., & de Beaufort, F. (1981). Protection, conservation de la nature et développement. In J. C. Lefeuvre, G. Long, & G. Riou (Eds.), *Ecologie et développement : Journées scientifiques* (pp. 31–98). CNRS.

Liu, M., Hu, Y., & Liao, M. (2009). Travelling theory in China: Contextualization, compromise, and combination. *Global Networks, 9*(4), 529–554. https://doi.org/10.1111/j.1471-0374.2009.00267.x

Payrot, J., Hartmann, V., & Cadène, F. (2014a). Plan de Gestion, Réserve Naturelle Marine de Cerbère Banyuls, Période 2015–2019, Section A – Diagnostic. Retrieved from https://www.ledepartement66.fr/wp-content/uploads/2021/06/PLAN-DE-GESTION-2015-2019_SECTION-A.pdf

Payrot, J., Hartmann, V., & Cadène, F. (2014b). Plan de Gestion, Réserve Naturelle Marine de Cerbère Banyuls, Période 2015–2019, Section B – Gestion. Retrieved from https://www.ledepartement66.fr/wp-content/uploads/2021/06/PLAN-DE-GESTION-2015-2019_SECTION-B.pdf

Préfecture Maritime Méditerranée. (2020). Arrête Préfectoral N° 040/ 2020, Règlementant la navigation, le mouillage, et al plongée sous-marine dans le périmètre de la Reserve Naturelle Marine de Cerbere-Banyuls. Retrieved from https://www.ledepartement66.fr/wp-content/uploads/2020/03/2020_040_AP_Reglementation-_navigation_mouillage_plongee_RNMCB.pdf

Rajaram, N., & Zararia, V. (2009). Translating women's human rights in a globalizing world: The spiral process in reducing gender injustice in Baroda, India. *Global Networks, 9*(4), 462–484. https://doi.org/10.1111/j.1471-0374.2009.00264.x

Snow, D. A. (2013a). Framing and social movements. In D. A. Snow, D. della Porta, B. Klandermans, & D. McAdam (Eds.), *The Wiley-Blackwell encyclopedia of social and political movements* (pp. 470–474). Malden, MA.

Snow, D. A. (2013b). Discursive fields. In D. A. Snow, D. della Porta, B. Klandermans, & D. McAdam (Eds.), *The Wiley-Blackwell encyclopedia of social and political movements* (pp. 367–371). Malden, MA.

Snow, D. A., & Benford, R. (1988). Ideology, frame resonance and participant mobilization. In B. Klandermans, H. Kriesi, & S. Tarrow (Eds.), *From structure to action: Comparing social movement research across cultures* (pp. 197–217). Jai Press.

Sunstein, C. R. (1996). Social norms and social roles. *Columbia Law Review, 96*(4), 903–968. https://doi.org/10.2307/1123430

Chapter 4
The Thau Fisheries Local Action Group

The Thau Fisheries Local Action Group (FLAG) is the case of focus in this chapter. After introducing the FLAG and identifying the local actor constellations involved in its governance and management, the chapter then identifies the specific global-level environmental norms brought to light in discourse surrounding this FLAG, providing an analysis of how they are adapted and translated to the local context. Showcasing mechanisms that help bridge the global with the local, this analysis highlights how local actors employ some of the dynamic processes discussed in Chap. 2 – including grafting, norm linkages, and normative reframing – in adapting global norms to the local setting, as well as how they draw upon specific global environmental norms in processes of frame construction that link environmental conservation with marine culture and activities.

Momentum surrounding the establishment of the Thau FLAG significantly increased in 2005 with efforts to preserve the Thau lagoon and its fishing and maritime farming activities (Syndicat Mixte du Bassin de Thau, 2017a). Located in the Occitanie region of France, the FLAG stretches along the Mediterranean Sea from Frontignan to Agde and includes two key fishing harbors in Agde and Sète. The 470 km^2 surface area of the Thau FLAG covers a population of over 150,000, including 2500 full-time fisheries employees spanning both fisheries and aquaculture (European Commission, 2017). In general, FLAGs are vehicles of Community-led Local Development (CLLD), an innovative "bottom-up approach to achieving territorial development in EU fisheries and aquaculture areas" (Miret-Pastor et al., 2020, p. 1). Through the actions of FLAGs such as the Thau FLAG, the CLLD approach aims to "increase both employment and territorial cohesion by bringing together local stakeholders in the selection and implementation of projects which meet the specific needs of the FLAG area and its fisheries communities" (Miret-Pastor et al., 2020, p. 1). Today, each of the European Union's (EU) five European Structural and Investment Funds (ESIFs), which are co-managed by the European Commission and EU member states, adopts the CLLD approach. This includes the

© The Author(s), under exclusive license to Springer Nature
Switzerland AG 2023
M. Schnyder, *Global Norms in Local Contexts*, SpringerBriefs in Political
Science, https://doi.org/10.1007/978-3-031-41108-3_4

European Maritime and Fisheries Fund (EMFF),[1] under whose organizational pur-
view FLAGs have implemented CLLD projects. Each FLAG has the authority to
create its own local development strategy, and to develop its own selection criteria
(based on local priority areas) to guide decisions about which projects to fund.[2]
There are five general categories of objectives for which CLLD projects can utilize
EMFF funds: (1) society and culture, (2) adding value to fisheries, (3) environment,
(4) governance and management, and (5) diversification (European Commission, n.d.).

The local development strategy of the Thau FLAG balances support for local
economic activities with consideration of the environmental assets and challenges
specific to the area, as challenges relating to water quality and sustainable management of land use have intensified over the past decade due to the increase in tourism
(European Commission, 2017). Challenges also exist in relation to the economic
viability of fishing activities in the area due to the strong seasonality of fishing practices, and new management procedures are needed for certain sensitive fish species
(European Commission, 2017). Given these and other challenges, the Thau FLAG's
strategy is defined by three strategic axes, as follows:

  – Creating local economic wealth by means of new economic activities and jobs, new
    production techniques, and new modes of governance and management,
  – Supporting fishery activities (jobs and professions) and local initiatives,
  – Sharing and pooling experiences and techniques, communicating to different stake-
    holder groups. (European Commission, 2017, 2)

An equal percentage of the Thau FLAG's operating budget is allocated across five
objectives: "strengthening the role of fisheries communities in local development;
promoting social wellbeing and cultural heritage; enhancing and capitalizing on the
environmental assets; supporting diversification; and adding value, creating jobs,
and promoting innovation along the fisheries chain" (European Commission, 2017,
2–3). Examples of some of the CLLD projects specific to the Thau FLAG include
developing methods of recycling fisheries waste, creating a shellfish observatory,
and establishing an oyster and wine route specific to the region (European
Commission, 2017, 3).

4.1   Thau FLAG Actor Constellations

The Thau FLAG involves 75 individuals and organizations working in partnership,
which includes a mix of the following stakeholders: public authorities (41%), fisheries actors (33%), private sector actors and NGOs (21%), and environmental
experts (5%) (European Commission, 2017). The FLAG is structured around four
main entities, to include a support structure in charge of its coordination and

---

[1] Now the European Maritime, Fisheries, and Aquaculture Fund.

[2] When selection is complete, the national-level managing authority of the EU member state in
which the FLAG is located approves projects as eligible for this funding (Miret-Pastor et al.,
2020, p. 2).

administrative management, a Technical Committee (CT) composed of representatives of the funders to provide technical and financial analysis of projects, a Selection and Steering Committee (CSP) in charge of selecting local projects and monitoring the implementation of the development strategy, and Thematic Working Groups (GTT) which are flexible structures that can be mobilized as necessary based on the priorities established in the development strategy, and can bring together local actors while providing supporting skills specific to the problems encountered (Syndicat Mixte du Bassin de Thau, n.d.-a). The FLAG's accountable body, or support structure, is the Syndicat Mixte du Bassin de Thau (SMBT),[3] where the Thau FLAG manager is based. The SMBT incorporates consultation bodies including the Joint Committee on which fishing and shellfish farming professionals sit, while the main decision-making body is the Syndicate Committee (Syndicat Mixte du Bassin de Thau, 2017b). The full FLAG partnership comprises 16 territorial collectivities (various regions, departments, and municipalities), 15 other public entities (tourist offices, chambers of commerce, etc.), 12 shellfish/aquaculture enterprises (including the Mediterranean Regional Shellfish Committee, aquaculture professionals, and shellfish producers unions), 12 fisheries entities (including the Regional Committee of fisheries and aquaculture, local Prud'homies, auctions, producer organizations, and other professionals), 4 scientific and educational organizations (including the French Institute for Ocean Science – Ifremer, the University of Montpellier, and maritime schools), and 16 representatives from other sectors (including environmental associations, marketing professionals, tour operators, event managers, etc.) (European Commission, 2017; Syndicat Mixte du Bassin de Thau, n.d.-b). As discussed above, these stakeholders collectively design and implement a local development strategy to address economic, social, and environmental challenges specific to the Thau FLAG's area, and select local projects to fund that contribute to development needs and priorities.

Many of the local stakeholders involved in the Thau FLAG can be viewed as both norm takers and norm translators. For example, those on the Technical Committee, Selection and Steering Committee, Syndicate Committee and other decision-making bodies often act as norm takers in the sense that they are situated on the receiving end of certain environmental norms being disseminated from the EU level of governance, given that the main objective of the FLAG's local development strategy is to realize Union Priority 4 of the EMFF – ensuring sustainable development in social, economic, and environmental areas (Marciano & Romeo, 2016). At the same time, other local stakeholders, such as those who are part of the consultation bodies, can be considered norm translators who apply and help integrate environmental norms into local projects, as decision-making surrounding the local development strategy is taken by a bottom-up approach that integrates stakeholders from the public, private, and civil society sectors. In this sense, actors involved in the Thau FLAG adapt the EMFF's funding guidelines to local circumstances and conditions specific to the FLAG area.

---

[3] In English, the Joint Syndicate of the Thau Bassin.

    Local actors involved in the FLAG's operation and administration, such as the
SMBT, can be viewed as norm entrepreneurs in that they aim to shape and make
more effective the norms that are applicable to local territorial development and
resilience. Often, these norms reinforce and reflect the broader global norm of sustainable development, which also aligns with EU-level priorities. Thus, it is likely
that those local actors engaged in supporting the Thau FLAG's activities are involved
in showcasing the coordinated EMFF framework based, in part, on the sustainable
development norm while at the same time working to independently adapt that
norm to the local territorial development context in ways that address the region's
challenges. Such approaches highlight the willingness by local actors to transmit
norms and rules from higher levels of governance, as well as to influence the evolution of regional-level sustainability norms in the Thau basin that focus on ecological
transition and strengthening territorial resilience. In a sense, the promotion of these
objectives also renders the SMBT, as well as those associated local actors involved
in the steering committee that selects local projects to fund, norm brokers in that
their work sometimes requires negotiating between the often-competing norms supporting sustainable development and those promoting the utilization of the Thau
lagoon to exploit its resources, to enjoy thermal waters, and to take advantage of the
other natural assets of the basin. As the SMBT states, it "has the task of ensuring the
maintenance and cohabitation of these traditional activities with other parts of the
economy of the basin" (Syndicat Mixte du Bassin de Thau, 2017c) – work that ultimately involves engaging with and navigating between alternative norms.

4.2   Global Environmental Norms

Which global environmental twilight norms are evident in the context of the Thau
FLAG? The goals of this section are, first, to detect those norms, and secondly to
highlight how specific global twilight norms are used in, and adapted to, the local
setting of the Thau FLAG area. Most of this section focuses on the sustainable
development norm, which features prominently in establishing a discursive field
surrounding the FLAG. The concepts of normative reframing and grafting, as well
as norm linkages, are showcased to illustrate how local stakeholders translate this
norm to the local setting. To a lesser extent, the global norm of equitable utilization
of shared natural resources is also present but occupies a far less prominent place in
discourse surrounding the FLAG.

4.2.1   Sustainable Development

The norm of *sustainable development* occupies a prime place in discourse related to
the Thau FLAG, which can be understood in part as a top-down norm being driven
from the EU level, as the primary aim of the FLAG's local development strategy for

the budget period 2014–2020 was to achieve the objective of Union Priority 4 of the EMFF as it relates to ensuring the sustainable development of its territory in social, economic, and environmental terms (Marciano & Romeo, 2016). This global norm has been heavily adopted into the strategic direction of the EMFF as its work was designed to contribute, in part, to the following Sustainable Development Goals (SDGs): SDG 1 End poverty in all its forms everywhere, SDG 2 End hunger, achieve food security and improved nutrition and promote sustainable agriculture, SDG 5 Achieve gender equality and empower all women and girls, and SDG 14 Conserve and sustainably use the oceans, seas and marine resources for sustainable development (European Maritime, Fisheries and Aquaculture Fund, 2022). At the EU level, sustainable development is further linked to the challenges that FLAGs in general are expected to face through 2030.[4]

As a result of its transmission from the EU level, the sustainable development norm features prominently in the themes and stated strategy of the Thau FLAG. Some of the themes supported by the EMFF, and by extension the FLAG, center on encouraging sustainable, innovative, and competitive fishing; encouraging sustainable, innovative, and competitive aquaculture; improving employment and strengthening territorial cohesion, and encouraging the marketing and processing of fishery and aquaculture products. Many of these themes are directly reflected in the Thau FLAG's strategy, which involves addressing challenges related to "creating and maintaining jobs and businesses in the fishery sector, [and] strengthening the importance of fishery and aquaculture activities within territorial development while ensuring sustainable development" (European Commission, 2017). Moreover, the Thau FLAG area has been characterized as having a "favorable local context" for CLLD initiatives to address these challenges, being distinguished by (among other things) "a strong dependence of certain municipalities on fishing and shellfish farming activities, whether they are exercised in the lagoon or at sea; a need to carry out innovative actions to strengthen the fisheries sectors, faced with major crises over the past ten years; [and] the existence of a dense web of potential partners and opportunities to implement these partnerships" (Syndicat Mixte du Bassin de Thau, n.d.-a, p. 6). The sustainable development norm also permeates the discourse of many of the FLAG's stakeholders, including the mission and work of the SMBT (the administrative and coordinating body of the FLAG), which describes itself as "a real lever for sustainable planning and development for the territory" (Syndicat Mixte du Bassin de Thau, 2017d). No other global environmental twilight norms feature as prominently as sustainable development in discourse specific to the Thau FLAG.

---

[4] Those are identified as: "sustainable food systems; climate change mitigation and adaptation; cleaner seas (including marine litter), balanced ecosystems and protection of marine biodiversity; developing business opportunities, including sustainable aquaculture and other blue growth sectors; a place for the young: within fisheries and the broader community; safe, quality jobs and social inclusion for all; a stronger role in governance and an improved image for fisheries" (Budzich-Tabor et al., 2020, p. 6).

With that said, it is worth briefly noting that the global norm of *equitable utilization of shared natural resources* is also detectable in the discourse of the SMBT. This twilight norm stipulates that when a specific natural resource forms a boundary between nation-states or is located across multiple countries, those states ought to cooperate in the management of the shared natural resource. Moreover, each of the nation-states would be entitled to an equitable share in the use of the resource. In this case, although the Thau basin is only located within one nation-state (France), the SMBT involves all of the communes of the Thau and Ingril watersheds, stretching across three intermunicipalities: Sète Agglopôle Méditerranée, the Community of Hérault Méditerranée Communes, and Montpellier Méditerranée Métropole (Syndicat Mixte du Bassin de Thau, 2017b). Therefore, it adopts a joint approach to the governance and management of the Thau ecosystem in which the principle of equitable utilization plays a role. To briefly illustrate, the SMBT states that "The joint preservation of this ecosystem and its economy requires support for the development of responsible and sustainable activities, controlled consumption of space, less polluting travel, truly protected nature and natural resources... A common goal that involves every one of us…" (Syndicat Mixte du Bassin de Thau, 2017h, n.p.). Under the principles of water management outlined by the SMBT, there is no place for "optimum" conservation, development, or use, since optimization would by definition have to occur at the expense of another involved actor. Instead, the operating principle of the SMBT reflects balanced conservation, development, and resource utilization, which captures the equitable utilization norm.

4.2.2   Norm Adaptation: Normative Reframing, Grafting, and Norm Linkages

The more prominent norm of sustainable development is adapted to the local context of the Thau FLAG region through norm linkages, normative reframing, and grafting processes. These processes are driven to a large extent by local fisheries actors – norm translators – who have proposed successful projects funded by the Thau FLAG. The project "Terre et Mer" provides a good example of these processes of adaptation that help integrate the global norm of sustainable development into the local context. Terre et Mer is a social enterprise project jointly undertaken by the local fish auction and the Red Cross that "links inclusion with the promotion of untapped local seafood" (van de Walle et al., 2019, p. 51). Motivated from the problem that 70% of the seafood being sold at the fish auction in Agde was being exported, the project set out to promote local consumption by using fresh, local seafood from the auction to process ready-made meals and ready-to-eat (e.g., tartare, carpaccio with marinade), ready-to-cook (e.g., fillet, gutted whole, etc.) packaged products (van de Walle et al., 2019, p. 51). While the local public can purchase the fresh production, which is marketed locally under a new brand, the frozen food is supplied to retirement homes and local school canteens (FARNET, 2020, n.p.).

The project's social enterprise component is reflected in the auction's collaboration with the Red Cross to "offer work to marginalised citizens under the supervision of a Production Manager and a Development Manager" (FARNET, 2020, n.p.) at a processing workshop in the auction building, where a total of 5 new jobs were created for the unemployed. A third component of the project involves linking seafood products and organic (sustainable) vegetable farming via a common trademark (FARNET, 2020, n.p.). Overall, the project "responds to consumer demand for ready-made meals while encouraging locals to eat seafood caught nearby" (van de Walle et al., 2019, p. 51), and has led to the creation of the first "social inclusion" seafood processing company in the Occitanie region. It promotes fish that are less well-known to consumers through the ready-to-cook and ready-to-eat product offerings, thereby adding value to these traditionally hard to sell varieties. At the same time, it promotes social inclusion and advocates for the unemployed by helping the unemployed develop skills and work experience via the processing workshop at the fish auction. Lastly, it promotes organic agriculture through its association of local seafood products and organic farming.

The Terre et Mer project is useful for illustrating the concept of normative reframing and how it can be used in processes of localization. Project stakeholders began by problematizing a specific situation – the fact that most of the catch was being exported – which helps to foreground at least two pre-existing norms that were considered problematic. The first is the norm that it does not matter who the fish is sold to, as long as it is being sold and fishers are earning an income. The second problematic norm is that local unemployment is a private matter specific to the individual experiencing it. Stakeholders foregrounded these status-quo norms; in other words, they called attention to them to showcase them as problematic and to call into question their normative "fit" to the issue at hand (Raymond et al., 2014). Through their project, they then proposed new norms to replace those that they deemed problematic, which centered on buying local (as opposed to exporting) and promoting social inclusion (as opposed to viewing local unemployment as a private matter).

In essence, stakeholders created linkages among these two norms that at first glance may appear unrelated. In fact, Béatrice Pary, the Thau FLAG Manager, has remarked that "This project is the result of an unprecedented collaboration between two worlds that still seem very distant: professional fishing and social inclusion" (quoted in van de Walle et al., 2019, p. 51). Indeed, one way that the broader sustainable development norm is integrated and adapted to the local context is via grafting it to the eat local and buy local norm, which can be seen in the promotion of local fish consumption (versus exporting). Through careful planning in view of stakeholder recognition of the growing demand for seafood products and for local, natural products (Syndicat Mixte du Bassin de Thau, n.d.-a), the norm translators involved in Terre et Mer framed the project this way, perfectly grafting onto existing local values, which helps to further socialize the local community to endorse norms related to sustainable development. Other projects financed by the Thau FLAG also graft sustainability to the norm of eating local, such as the Conch en Mer project, which was launched to examine the economic and financial viability of a shellfish farming project in open sea production areas on the coast, with one of the main goals being to reclaim the open-sea mussel market in order to strengthen the position of regional production in a market highly dependent on imports (DLAL FEAMP, n.d., n.p.). Prior research has documented how grafting novel (global) norms to previously accepted (local) norms often facilitates acceptance and strengthens the norm in question (Price, 1998; Acharya, 2004).

Furthermore, the two seemingly unrelated norms of eating/buying local and promoting social inclusion are linked under the broader sustainability norm supporting short supply chains. Indeed, the concept of short supply chains plays a key role in how the sustainable development norm is adapted to the local context of the Thau FLAG, and is reflected in the long-term territorial development strategy of the region (Syndicat Mixte du Bassin de Thau, n.d.-c). For example, publicity about the project has featured its creation of short supply chains to emphasize the promotion of good products caught locally, benefitting the local economy (Scheffer, 2020). Under the Terr'Iodée brand, the project has publicized how it sells in short supply chains to local shops, restaurants, and even primary school canteens (Scheffer, 2020). In addition, the social integration component of the project is presented as integral to relying on short supply chains to process the fish, which also produces benefits to local fishermen (Midi Libre, 2020).

The Seafood Baskets project (Paniers de Thau) of the Thau FLAG, launched in 2012, is another notable example of how grafting and normative reframing processes are used in norm adaptation. The Centre Permanent d'Initiatives pour l'Environnement – a local environmental association – developed the seafood basket distribution scheme to help connect local fishermen and aquaculture producers to local consumers, with the aim of developing a market for local fisheries products. Although inhabitants of the villages around the Thau lagoon live in a region with a strong fisheries tradition, this initiative stemmed from the problem that they nonetheless had difficulties when it came to buying local seafood. As a result of there being no local fishmongers, local inhabitants needed to track down individual producers willing to sell their products or else purchase products from larger stores (FARNET, 2014). The issue of there not being enough producers involved in short supply chains together with the irregular supply of products are problems that were actively foregrounded as weaknesses of the Thau region by local stakeholders involved in managing the Thau FLAG (Syndicat Mixte du Bassin de Thau, n.d.-a, p. 17). In this regard, the real norm being foregrounded as problematic was not buying local, as consumer practices were contradicting the region's heavy emphasis on short supply chains and support for the local marine cultural heritage. In fact, stakeholders were confronting a real "lack of knowledge about local products (by residents and visitors)" (Syndicat Mixte du Bassin de Thau, n.d.-a, p. 16). This is, in part, reflected in the fact that the project started from a single direct sales initiative, evolving over time "into a distribution network active in four villages, backed up by a website enabling customers to register for the scheme, place orders and select their pick-up point" (FARNET, 2014, p. 1), based on principles similar to those of Community Supported Fisheries (van de Walle et al., 2019, p. 30).

To replace the foregrounded norm that was supporting the problematic situation, stakeholders positioned the initiative as promoting short supply chains, as they had already identified a trend toward growing demand for this among consumers (Syndicat Mixte du Bassin de Thau, n.d.-a). Similar to the Terre et Mer project, this trend enabled stakeholders to graft the broader norm of sustainable development onto the local normative framework supporting local, natural products, reflected in the promotion and planning of direct sales either on the production sites, or via "counters of the seas" that sell local seafood (Syndicat Mixte du Bassin de Thau, n.d.-a, p. 20). Over time, the baskets have evolved to combine seafood offerings with other local products including bread, cheese, and other local delicacies, enabling the baskets to become "a true socialising tool for the area" (FARNET, 2014, p. 1).

4.3   A Matter of Preserving and Promoting Cultural Heritage

In the Thau basin, stakeholder discourse focuses heavily on *the ecological transition*. The ecological transition is a normative idea related to sustainable development that guides policy and involves the creation of innovative tools to design and manage the territory over the long term (10–30 years), with some of the objectives being the promotion of a pleasant living environment, preserved nature, resilience in the face of environmental risks and challenges, a focus on innovation while being resource-efficient, and an ability to provide jobs while being equipped with the appropriate infrastructure (Syndicat Mixte du Bassin de Thau, 2017e). Principles supporting the ecological transition are reflected in the Territorial Coherence Scheme (SCOT), which lays out the major objectives for the region in setting the strategy and framework for the planning and development of the territory over the next 2–3 decades. Revised in 2017, the current SCOT sets as priorities the acceleration of the ecological transition and strengthening the resilience of the territory (Syndicat Mixte du Bassin de Thau, 2017f).

In light of the regional planning principles and objectives laid out in the SCOT, how is frame construction used in the local context of the Thau FLAG region to adapt the global norm of sustainable development? In short, local stakeholders construct frames that position the value of sustainable development as intrinsically linked to the preservation and promotion of the local marine cultural heritage. The concept of "patrimoine" (heritage) is frequently invoked as a way to frame the norm of sustainable development as a local matter of critical importance. The SMBT (the FLAG's administrative body), for example, states that "Fishing and shellfish farming are organized around the Thau lagoon, which is an exceptional environmental, social and economic part of our heritage. To preserve these activities on Thau, it is essential to ensure the preservation of water…and the lagoon. The Thau territory depends on the outside for 80% of its drinking water supply. It is necessary to ensure good management of this water resource so that in years to come, in 10, 20, 30 years, all the inhabitants will have enough water for drinking, washing, watering

their garden, irrigating their crops, supplying their industries, businesses..."
(Syndicat Mixte du Bassin de Thau, 2016, p. 3). In this sense, preserving the Thau
is framed as being explicitly linked to preserving the local heritage. The seafoods
baskets project funded by the Thau FLAG also employs the term *heritage*, and uses
similar framing that connects the sustainable development idea that motivated this
project to the promotion of the local fisheries heritage as cultural heritage: "Enabling
meet-and-greets and informal exchanges between customers and professionals
around their products, the scheme is now diversifying its offer towards local 'cultural' products connected with the fisheries heritage such as recipe contests, arts and
crafts and theatre performance" (FARNET, 2014, p. 1). The concept of *patrimoine*
is therefore an important framing device that helps explain how the global sustainable development norm is translated to this local coastal environment.

Furthermore, the SMBT regularly communicates messages that forge a strong
linkage between sustainable development and the local marine culture of the Thau
region. This general idea can be seen, for instance, in community-led local development strategy reports (e.g., Syndicat Mixte du Bassin de Thau, n.d.-a, p. 23). An
illustration of how the sustainable development norm is linked to the Thau region's
marine culture is evident in the following excerpt from the SMBT, in which it conceptualizes the economic future of the region:

> The exceptional environmental assets of the Thau basin can boost the economy. The terri-
> tory already relies on well-established strengths: the blue economy (fishing, marine culture,
> yachting, maritime trade, seaside tourism), hydrotherapy, agriculture, eco-tourism... Many
> factors can influence the maintenance or development of these activities such as the reserva-
> tion of dedicated spaces and resources, environmental protection or even the control of
> attendance. On Thau, economic development must be sustainable, inclusive and resilient.
>
> Add to that a strong agricultural identity and a few hundred thousand tourists a year and
> you get the portrait of an atypical local economic fabric. This ecosystem is largely oriented
> towards the coast and its assets. It is made up of activities that cannot be relocated and for
> the most part are in permanent interaction with the natural environment. Preserving it and
> helping it to develop calls for constant monitoring because this economic fabric depends on
> the quality of an environment that is constantly in demand and on ever stricter regulations.
> And if it is essential for each of these traditional sectors to orient its model towards prac-
> tices that respect the resource, it is also essential that everyone has their own dedicated
> space with an appropriate level of equipment and infrastructure. (Syndicat Mixte du Bassin
> de Thau, 2017g, n.p.)

In essence, the framing reflects the positioning that the growth of the economy of
the Thau basin must be sustainable and respectful of natural resources, in part
because the economy is so strongly tied to those resources. Because of this, sustainable development in the context of the Thau region is intrinsically linked to the
preservation and promotion of the area's marine cultural heritage.

In addition, sustainable development is positioned as a matter of maintaining
connections between land and sea. In other words, the issue of sustainable development in land-based agriculture is positioned as being tied to the blue economy of the
area. To promote biodiversity and encourage short supply chains, the SMBT
explains that maintaining the agricultural areas of the Thau region and encouraging
the installation of new farmers involves promoting the link between land and sea,

both in the promotion of products and in the establishment of short supply chains (Syndicat Mixte du Bassin de Thau, 2017i). From this lens, the issue of sustainable economic development also necessitates forging new and innovative relationships between farmers and fishermen.

4.4 Conclusion

This chapter has examined the case of the Thau Fisheries Local Action Group, located along the Mediterranean Sea in France's Occitanie region. After providing a brief overview of the FLAG and identifying the constellations of local stakeholders involved in its governance and management, the main focus of the chapter centered on identifying the specific global environmental norms evident in discourse specific to this FLAG by local stakeholders, and on assessing how those norms are being adapted to the local context. The examples analyzed throughout the chapter have showcased how local actors employ processes related to norm grafting, norm linkages, and normative reframing. The chapter also examined the construction of issue frames that link environmental conservation with the marine activities and culture of the Thau region.

This analysis of the Thau FLAG has produced several findings worth noting. First, the global norm of sustainable development appears to be the most prominent norm reflected in stakeholder discourse surrounding the FLAG. Although other global norms, such as the equitable utilization of shared natural resources, are identifiable in terms of the long-term governance and management of the Thau basin, these are not the predominant norms that stakeholders draw upon in the context of the management and governance of the Thau FLAG itself. In addition, the Terre et Mer project and the Seafood Baskets project illustrate how Thau FLAG actors use normative reframing to call into question norms deemed undesirable, while working to replace them with other norms focused on eating and buying local that better reflect the principles of sustainable development that are consistent with the long-term territorial development strategy of the region. The projects also show how stakeholders are using the process of grafting to translate the broader sustainable development norm to the local context. Recognizing the growing demand in the area for seafood products and for local products, the norm translators involved in Terre et Mer and Seafood Baskets framed the project in a way that grafts sustainable development principles onto existing local values that emphasize the importance of local products and short supply chains. In this way, these examples help illustrate how some of the norm-based strategies that are typically showcased in the literature on institutional change (Raymond, 2016; Raymond et al., 2014) can also play a role in processes of norm localization. Lastly, taking advantage of sustainability principles articulated in the long-term development strategy of the region, norm-makers in the Terre et Mer project use norm linkages to connect the two seemingly unrelated norms of eating/buying local and social inclusion under the broader normative framework supporting short supply chains.

The examples examined in this chapter have also shed light on issue frames and frame construction, highlighting how local norm entrepreneurs and norm translators position the importance of sustainable development as part and parcel of the preservation of the maritime cultural heritage specific to the Thau region. Just as in the case of the Cerbère-Banyuls MNR examined in the last chapter, norm-makers in the Thau FLAG draw upon the concept of *patrimoine* (heritage) in positioning sustainable development as a local matter with high stakes, which involves preservation in both a cultural and economic sense. The way that certain global twilight norms are framed using the *patrimoine* concept represents an important process of norm localization. Overall, this chapter underscores a central argument from the norm localization literature: that norm translation involves complex processes whereby local actors build congruence between the global norm and local values. In this sense, the examples presented here do not lend support to the view that we should expect a static fit between global norms and local values (Checkel, 1999; Young, 1999; Cardenas, 2007); rather, congruence processes are undertaken by local actors – "inside proponents" (Acharya, 2004) – who utilize dynamic acts to adapt the global norm to the local environment in ways that resonate with the ideas and values that are important in that specific context.

References

Acharya, A. (2004). How ideas spread: Whose norms matter? Norm localization and institutional change in Asian regionalism. *International Organization, 58*(2), 239–275. https://doi.org/10.1017/S0020818304582024

Budzich-Tabor, U., Vercruysse, J.-P., van de Walle, G., Posti, J., Rigaud, A., & Veronesi Burch, M. (2020). *Forward-looking strategies for fisheries areas, Farnet guide #20*. European Union.

Cardenas, S. (2007). *Conflict and compliance: State responses to international human rights pressure*. University of Pennsylvania Press.

Checkel, J. T. (1999). Norms, institutions, and national identity in contemporary Europe. *International Studies Quarterly, 43*(1), 83–114. https://doi.org/10.1111/0020-8833.00112

DLAL FEAMP. (n.d.). Conch en mer. Retrieved from https://www.dlalfeamp.fr/projet/conch-en-mer/

European Commission. (2017). *FLAG factsheet: Thau Flag*. Retrieved from https://webgate.ec.europa.eu/fpfis/cms/farnet2/on-the-ground/flag-factsheets/thau-flag_en.html

European Commission. (n.d.). FARNET themes. Retrieved from https://webgate.ec.europa.eu/fpfis/cms/farnet2/themes_en.html

European Maritime, Fisheries and Aquaculture Fund. (2022). Programme in a nutshell: Program statement. Retrieved from https://commission.europa.eu/strategy-and-policy/eu-budget/performance-and-reporting/programme-performance-overview/european-maritime-fisheries-and-aquaculture-fund-performance_en#programme-statement

FARNET. (2014). *Project summary #041-FR09-EN – Seafood baskets*. European Commission.

FARNET. (2020). *Good practice project: A social enterprise that places local fish in school canteens*. Retrieved from https://webgate.ec.europa.eu/fpfis/cms/farnet2/on-the-ground/good-practice/projects/social-enterprise-places-local-fish-school-canteens_en.html

Marciano, C., & Romeo, G. (2016). Integrated local development in coastal areas: The case of the "Stretto" coast FLAG in southern Italy. *Procedia - Social and Behavioral Sciences, 223*, 379–385. https://doi.org/10.1016/j.sbspro.2016.05.251

Midi Libre. (2020). *L'entreprise "Brise de Terre" mise sur le circuit court pour transformer les poissons*. 13 November. Retrieved from https://www.midilibre.fr/2020/11/13/a-la-criee-dagde-linsertion-sociale-permet-aussi-daider-les-pecheurs-9197255.php

Miret-Pastor, L., Svels, K., & Freeman, R. (2020). Towards territorial development in fisheries areas: A typology of projects funded by fisheries local action groups. *Marine Policy, 119*, 104111. https://doi.org/10.1016/j.marpol.2020.104111

Price, R. (1998). Reversing the gun sights: Transnational civil society targets land mines. *International Organization, 52*(3), 613–644. https://doi.org/10.1162/002081898550671

Raymond, L. (2016). *Reclaiming the atmospheric commons: The regional greenhouse gas initiative and a new model of emissions trading*. MIT Press.

Raymond, L., Weldon, S. L., Kelly, D., Arriaga, X. B., & Clark, A. M. (2014). Making change: Norm-based strategies for institutional change to address intractable problems. *Political Research Quarterly, 67*(1), 197–211. https://doi.org/10.1177/1065912913510786

Scheffer, H. (2020). Brise de terre : l'insertion en circuit court à Agde. *Produits de la Mer*, 14 April. Retrieved from https://www.pdm-seafoodmag.com/lactualite/brise-de-terre-linsertion-en-circuit-court-a-agde/

Syndicat Mixte du Bassin de Thau. (2016). Le sage qu'est-ce que c'est ? *Le Journal des P'tits Sages de Thau, 1*, 3. Retrieved from https://www.smbt.fr/blog/outil/le-sage-des-enfants/

Syndicat Mixte du Bassin de Thau. (2017a). Un peu d'histoire. Retrieved from https://www.smbt.fr/le-smbt/un-peu-dhistoire/

Syndicat Mixte du Bassin de Thau. (2017b). Le SMBT en bref. Retrieved from https://www.smbt.fr/le-smbt/le-smbt-en-bref/

Syndicat Mixte du Bassin de Thau. (2017c). Favoriser l'économie bleue du littoral. Retrieved from https://www.smbt.fr/blog/mission/favoriser-leconomie-bleue-du-littoral/

Syndicat Mixte du Bassin de Thau. (2017d). Qu'est-ce que le Syndicat mixte du bassin de Thau? Retrieved from https://www.smbt.fr/

Syndicat Mixte du Bassin de Thau. (2017e). Aménager un territoire solidaire et durable. Retrieved from https://www.smbt.fr/blog/mission/amenager-un-territoire-solidaire-et-durable/

Syndicat Mixte du Bassin de Thau. (2017f). La cohérence territoriale (SCOT). Retrieved from https://www.smbt.fr/blog/outil/scot/

Syndicat Mixte du Bassin de Thau. (2017g). Notre avenir. Retrieved from https://www.smbt.fr/blog/enjeu/votre-avenir/

Syndicat Mixte du Bassin de Thau. (2017h). Notre territoire. Retrieved from https://www.smbt.fr/blog/enjeu/notre-territoire/

Syndicat Mixte du Bassin de Thau. (2017i). Accompagner les pratiques agricoles durables. Retrieved from https://www.smbt.fr/blog/mission/favoriser-leconomie-bleue-du-littoral/accompagner-les-pratiques-agricoles-durables/

Syndicat Mixte du Bassin de Thau. (n.d.-a). DLAL FEAMP 2014–2020: Thau et sa bande côtière de Frontignan à Agde: Dossier de candidature actualisé 2017. Retrieved from https://www.dlalbassindethau.fr/wp-content/uploads/sites/3/2017/05/2017-Candidature-DLAL-Thau.pdf

Syndicat Mixte du Bassin de Thau. (n.d.-b). Le groupe DLAL. Retrieved from https://www.dlal-bassindethau.fr/le-groupe-dlal/

Syndicat Mixte du Bassin de Thau. (n.d.-c). Schéma de cohérence territoriale de Bassin de Thau : Document d'orientations et d'objectifs. Retrieved from https://www.smbt.fr/blog/outil/scot/

van de Walle, G., Veronesi Burch, M., & De Smet, S. (2019). *Post 2020: Local action in a changing world*. European Commission, Directorate-General for Maritime Affairs and Fisheries, Director-General.

Young, O. R. (1999). Regime effectiveness: Taking stock. In O. R. Young (Ed.), *The effectiveness of international environmental regimes: Causal connections and behavioral mechanisms* (pp. 249–280). MIT Press.

Chapter 5
The Biovallée Biodistrict

The final case – the Biovallée biodistrict – is examined in this chapter. Similar to the structure of the previous two chapters, Chap. 5 begins by introducing the Biovallée biodistrict, providing a brief history, and identifying the actor constellations involved in its governance and management. From here, it identifies the specific global-level environmental twilight norms that are evident in the discourse of local stakeholders and source documents specific to this case. Like the previous two chapters, the analysis illustrates theoretical processes and constructs discussed in Chap. 2 in showcasing the mechanisms by which local stakeholders have adapted and translated global norms to the local context of the Biovallée. Specifically, this chapter highlights the novel processes of foregrounding and normative innovation, along with the creation of new concepts that help bridge the global to the local.

Biodistricts are territories "where farmers, citizens, associations and public authorities enter into an agreement for the sustainable management of local resources" and are considered participatory and innovative multi-stakeholder governance arrangements (Assaël, 2017, p. 1.). Based on organic agricultural methods and techniques, they represent a new way to sustainably reconvert entire territories. In other words, biodistricts represent a local strategy for sustainable territorial development. The biodistrict's integrated approach to territorial development means that "the promotion of organic products is closely linked with the promotion of the territory, to achieve the full development of its economic, social and cultural potential" (Assaël, 2017, p. 1.). Economically, the aim is to produce more market opportunities for organic producers while reducing input costs and increasing scale; environmentally, the aims center on increasing local biodiversity, and making agriculture more sustainable while reducing its environmental impact; and socially, the aims include promoting rural employment and land access for younger generations, enhancing knowledge exchange between different stakeholders, and promoting food sovereignty and local food cultures (Basile, 2017).

M. Schnyder, *Global Norms in Local Contexts*, SpringerBriefs in Political Science, https://doi.org/10.1007/978-3-031-41108-3_5

In 2004, the Italian Association for Organic Agriculture (AIAB) began work to promote the launch of the first biodistrict in Cilento (Salerno), Italy, which initially involved organizing forums and public meetings with agricultural associations, municipalities, and other local actors interested in developing a new framework for sustainable resource management (Basile & Cuoco, n.d., p. 8). From these meetings, the concept of the biodistrict was created. Initially, the push to establish a biodistrict came from a group of local organic farmers who wanted the local authorities to help develop a local organic market (Basile, 2017, p. 2). In 2009, the AIAB officially launched the first biodistrict (Basile, 2017), which was then formalized into a governance structure with the signing of an agreement between the different municipalities and an act passed by the Campania Region (Basile & Cuoco, n.d., p. 8). The goal for those involved centered on undertaking joint actions for inclusive territorial development strategies, which consisted of "pilot organic group certification, setting up production guidelines for farmers, [an] awareness campaign of sustainable agriculture production along with public procurement schemes to promote [the] Mediterranean diet and local organic consumption in schools and hospital canteens, and local administration offices" (Basile, 2017, p. 2). The Cilento biodistrict covers an area of 3196 square kilometers, covering 38 municipalities and 400 organic farms (Bio-distretto Cilento Association, 2019, p. 2). As of 2020, there were 48 established biodistricts in Europe, primarily clustered within Italy and surrounding areas, and an additional ten in development in Italy and Spain (EducEcoRegions Project, 2020).

Currently, the Biovallée is the only biodistrict in France. It is located in the Drôme valley, a rural area of 2200 km^2 in the Rhône-Alpes region in the southeast of France, where organic production represents 30% of agricultural land (well above the national average of 3%) (Bui & Lamine, 2015a). The valley's geographical territory consists of the Drôme river's watershed and surrounding mountains, comprising 102 small towns and villages with a population of approximately 54,000 inhabitants (Biovallée n.d.-a). The reach of the Biovallée involves two districts with influence on the project: an upstream district (Diois) and a downstream district (Val de Drôme).[1]

In 2009, the 102 municipalities of the valley launched the Biovallée biodistrict as a public project, which was formally approved by the Regional Council in the same year (Biovallée, 2014). However, its roots can be traced back to the emergence of organic production in the valley in the 1970s (Biovallée, n.d.-b). Today, the broad aim of the Biovallée project is to make the Drôme a regional leader in the preservation and valuation of natural resources. To this end, it has enumerated many objectives, some of which include developing high-level training in the field of sustainable development, supplying 25% of the territory's energy consumption through local production of renewable energy by 2025 and 100% by 2040, supplying 80% of the procurement of institutional catering with organic or regional products by 2025,

---

[1] In reality, there are three districts involved in the Biovallée project, however only the two biggest, representing 88 municipalities, have an influence (Bui & Lamine, 2015a).

developing training and research related to sustainable development (10 partner-ships in 2012, 25 targeted in 2025), and creating 2500 new jobs in the territory in eco-sectors by 2025 (Biovallée, n.d.-c).

5.1 Actor Constellations in the Biovallée Biodistrict

As is the case with all biodistricts, there are six main sets of stakeholders involved in the Biovallée biodistrict. At the heart of the biodistrict are farmers (producers). They are the major stakeholders, adhering to the rules of organic farming. The organic farmers receive distinct benefits from the biodistrict, including access to a local market for their produce, being part of a multifunctional tourist circuit, and promoting their produce in the context of the biodistrict's territorial marketing plans (Basile & Cuoco, n.d.). In the Drôme valley, approximately 15% of the population in the upper valley were farmers around the time of the biodistrict's creation (of which 25% were already practicing organic production), compared to only 2.5% of the lower valley's population. However, as local councils became more interested in organic agro-industries and their potential in terms of local development, they sug-gested a cross-valley collaboration to take advantage of the unique strengths of the upper and lower valley (IPES-Food, 2018).

As other researchers have noted, the Biovallée case "provides insights into how norms can be shifted over time" as a result of "ongoing interaction between main-stream and alternative actors" (IPES-Food, 2018, p. 59). Over time, organic farmers became key norm entrepreneurs able to spread the vision for a sustainable agricul-tural model to conventional farmers via shared spaces. Peer-to-peer knowledge sharing networks, including agricultural knowledge exchange groups like the Centre d'études techniques agricoles (or CETAs), provided a critical space for contact and interaction between organic and conventional farmers, which ultimately helped bring about changes in the norms driving production practices (IPES-Food, 2018). Norm entrepreneurs interested in changing production norms to better align with the values of a number of local farmers seeking alternatives (in part to lessen their reliance on external inputs) successfully used these channels, where "the logic of organic agriculture was progressively shared, legitimized, and mainstreamed" (IPES-Food, 2018, p. 62). As a result, organic production became incorporated into the local agricultural landscape as opposed to remaining a niche. The Biovallée project has helped to institutionalize these norms.

Consumers represent another stakeholder set. Citizens in a biodistrict benefit by being able to purchase local organic products through short supply chains, and from the increased environmental quality that comes from organic farming. In addition, consumers can establish direct relations with the producers, building trust and fos-tering mutual cooperation. In the Biovallée, La Carline, which began as a small organic buying group consisting of a few households in the Drôme Valley, had grown to 600 participating households by 2008 (IPES-Food, 2018). As consumers interact with organic farmers, they can be thought of as norm-takers, or recipients of the novel norms that farmers have helped to create in the biodistrict context. Consumers can also help to spread or transmit existing norms through their discourse with other consumers.

In addition, public authorities are critical stakeholders. Local and public authorities act as norm translators in the sense that they commit to spreading organic agriculture and food culture through various initiatives. They also jointly organize biodistrict events and apply sustainability principles to other planning and economic activities. In the Biovallée, governance is primarily de facto shared between the elected officials from the upstream and downstream districts. However, the governance process is structured to take into account the opinions of the technical committee (which consists of representatives from both districts, the Chamber of Agriculture, local farmers' organizations, and local agricultural training organizations) and of the two districts' commissions on agriculture.[2] In practice, other local actors are represented in intermediary stages but not in final decision-making processes of the Biovallée (Bui & Lamine, 2015a). Yet at the level of the upper and lower districts themselves, governance is shared among all local actors. For instance, local actors (including agricultural development and training organizations, farmers' unions, farmers' cooperatives, and civil society) participate within the two districts' commissions on agriculture (Bui & Lamine, 2015a).

Industry in the agro-food supply chain represents another set of major stakeholders in the biodistrict. Both the food industry and the agricultural equipment industry can benefit from the high concentration of organic farms in the area in terms of supplying agricultural equipment and in terms of the production of raw materials for food processing (Bui & Lamine, 2015a). The tourism and catering industries are also major players, benefitting from expanded product offerings (organic menus) and agro-tourism (organic trails, farm visits, etc.) (Bio-distretto Cilento Association, 2019).

Associations are the fifth main set of actors. Environmental, agricultural, ecotourism, and other associations work individually or in conjunction with other stakeholders to promote activities within the biodistrict. For example, the Agricourt association was established in the Biovallée to help facilitate local sustainable sourcing for cafeterias and private day care centers (IPES-Food, 2018). Finally, local training and research centers are relevant actors. They organize trials and training initiatives to enhance the single initiatives of local stakeholders, and in the Biovallée, they participate in governance via membership on the technical committee (Bui & Lamine, 2015a).

---

[2] Although the Biovallée territory actually consists of three districts (Val de Drôme, Crestois Pays de Saillans, and Diois) (Biovallée, 2019), the largest two districts (Val de Drôme and Diois) dominate decision-making.

5.2   Global Twilight Norms

To what extent are any of the global environmental twilight norms apparent in the context of the Biovallée biodistrict? The goals of this section are twofold: first, to identify those norms and where they appear, and secondly, to showcase how they are translated to the local setting of the Biovallée biodistrict. The sustainable development norm, which was prominent in the cases of focus in the last two chapters, also features prominently in the discursive field surrounding the biodistrict. The intergenerational equity norm is also evident, albeit to a lesser extent. Following the identification and discussion of these twilight norms, the concept of normative innovation is introduced to showcase how local stakeholders translate the sustainable development norm to the local context.

5.2.1   Sustainable Development and Intergenerational Equity

The sustainable development norm appears throughout a multitude of Biovallée documents that explain the development of the biodistrict and articulate its objectives. This norm tends to be referenced in two particular contexts: in discussing the aims of the biodistrict, and in explaining how its existence sets a regional and global standard while transmitting a culture of sustainability. For example, in explaining the aims of the Biovallée, a case study from the French National Institute for Agriculture, Food, and Environment (INRAE) focuses on its historical development, stating that "It builds upon the long experience of local public policies focused on rural development from the 1970s and 1980s and on sustainable development since the 1990s. This Biovallée project aims at making the Drôme valley a pilot territory in terms of sustainable development" (Bui & Lamine, 2015b, p. 1). A report on organic districts in Europe echoes this norm in elaborating the biodistrict's aims: "The main aim is to create a global sustainable food system, able to ensure healthy and sustainable diets for all. In accordance, of course, with the local diversity and products," (EducEcoRegions Project, 2020, p. 9). It goes on to directly address the global-level norm: "The experience of Organic Districts has shown how it is possible, in a relatively short time, trying [sic] to make territories better, while contributing to the pursuit of the United Nation's [sic] sustainable development objectives" (EducEcoRegions Project, 2020, p. 10). In short, there is much discourse echoing the theme that the Biovallée project aims to make the Drôme valley an example of success for the sustainable preservation and enhancement of natural resources (Biovallée, n.d.-d; Bui & Lamine, 2015a), by way of specific target goals to achieve energy autonomy, transform agricultural and food practices, change consumption habits, and reduce waste.

The sustainable development norm is also referenced in relation to how the Biovallée establishes a particular benchmark and spreads the values of organic agriculture on a wider scale. For instance, an evaluation created for the steering committee articulates how the Biovallée sets a regional standard, seeking to represent "a national benchmark for sustainable development," (Biovallée, 2014, p. 1) and goes on to enumerate specific benchmarks that set the standard for other regions, which include developing high-level training in the field of sustainable development, and undertaking research related to sustainable development (Biovallée, 2014). An assessment of the protocol by budget line shows specific funds geared toward pragmatic actions that advance communities, associations, companies, and inhabitants of the valley "in their practices of sustainable development" (Biovallée, 2014, p. 4), while "the vision of the future it proposes in its slogan, 'an exemplary territory [or a territory school] for sustainable development,' reflects the global approach of the Biovallée program" (Bui & Lamine, 2015a, p. 16). These illustrations highlight how the Biovallée itself can act as a norm transmitter by actively pursuing the diffusion of sustainable development practices and principles at various scales and levels.

In this respect, the biodistrict seeks to transmit a broader culture of sustainability to other regions and territories. This is illustrated, in part, by how the Biovallée represents a "symbol of a global approach to 'sustainable development' (in reference to the Brundtland report)" (Biovallée n.d.-e, n.p.), and also by the commitment that territories make to the ideals of the Biovallée. That is to say, when "communities of municipalities implement exemplary projects" related to the biodistrict, "the culture of sustainable development is gradually being extended to all their services" (Biovallée, 2014, p. 18). It is this idea of the cultural spread of the sustainable development norm that positions the biodistrict as a norm transmitter.

To a lesser extent, the norm of intergenerational equity is also present in the discourse of the Biovallée. For example, the biodistrict has been credited with promoting an "ecological transition and ensur[ing] a sustainable and fair-trade future for the new generations," while advancing sustainability-related policies at the international and European levels (EducEcoRegions Project, 2020, p. 9). This connects to the intergenerational equity norm in its emphasis on using the environment and natural resources in a way that ensures their preservation for the benefit of generations to come. Further, a report by the National Institute of Agricultural Research (INRA) in France has noted "the interdependency of stakeholders and dimensions and their role in the future (social, economic and ecological) resilience of farms, food system, and territory at large" (Bui & Lamine, 2015a, p. 20), which is suggestive of the intergenerational equity norm in its focus on the preservation of the land and territory for the future. The Biovallée website further reinforces this emphasis on intergenerational equity and places it in the context of a continuation of local resistance. Speaking of how the Drôme valley was transformed at the beginning of the twentieth century by waves of new arrivals, it speaks of the residents, businesses, communities, and associations involved in the biodistrict:

> These people and structures continue the collective actions, the actions of resistance carried out during the previous centuries to protect, to preserve the commons: the forests, the rivers, the air and the soil. It is on this basis that the Biovallée takes shape. A project to achieve energy autonomy, transform our agricultural and food practices, our consumption habits or even reduce our waste and travel. A global approach to developing a local and sustainable economy where new ways of living are invented that preserve natural resources and ensure access to essential services for all inhabitants. (Biovallée, n.d.-d, n.p.)

It is through this idea of local resistance to the mainstream that the process of normative innovation is reflected – a norm-based process which Biovallée stakeholders employ to localize the global norms that are evident in discourse surrounding it. The next section unpacks and explains this process of norm creation and how stakeholders have used it to adapt global norms to the local context.

5.3 Norm Adaptation Through Foregrounding and Normative Innovation

Recall from Chap. 2 that normative innovation is a strategy for bringing about social or political change that entails introducing a novel norm in relation to a particular situation. Before a novel norm can be introduced, however, stakeholders must weaken the existing normative framework via a process of foregrounding. Through foregrounding, stakeholders call attention to a problematic pre-existing norm by emphasizing its harmful impact, which, in turn, prompts others to reconsider the norm's implications (Raymond et al., 2014). In this way, they work to undermine the normative status quo by weakening its appeal. Stakeholders then introduce a new, alternative norm to prompt new behaviors, action, or policies.

In the context of the Biovallée, stakeholders have called into question the norms and behaviors that uphold the mainstream, status quo model of economic development. For example, as a biodistrict, the Biovallée is positioned as "a concrete response to the current trend of economic development, causing massive phenomena of abandonment of rural areas and the growing urbanization of people in search of better living conditions and higher incomes. The process affects both the most industrialized and developing countries, causing the degradation and progressive impoverishment of local resources, the loss of biodiversity and the traditional knowledge of local cultures" (EducEcoRegions Project, 2020, p. 9). Here, several concrete problems associated with the dominant economic model are highlighted, including rural flight, along with the degradation of the environment and the pervasive loss of local cultural knowledge and traditions. Local actors have actively called attention to the harms associated with an internationalized food supply chain that transports products across continents at a heavy environmental cost associated with road and air transport. It is the norms, or the standards of behavior, that support this dominant model that stakeholders seek to highlight and weaken.

Prior to the establishment of the Biovallée, the spread of organic farming in the Drôme area in the 1970s took place in a context strongly dominated by the agricultural modernization paradigm, with most farmers adhering to the dominant industrial model of agriculture. Yet as research into the history of the Biovallée has documented, "this more conventional approach jarred with the values of a number of local farmers seeking alternatives…" (IPES-Food, 2018, pp. 61–62). It is precisely such a clash of values and norms that sets the stage for stakeholders to engage in normative innovation. The industrial model of production and the values it rested upon were being foregrounded for quite some time prior to the formal establishment of the Biovallée. This process of foregrounding what were considered to be undesirable standards of behavior and practice involved "decades of efforts by local actors not only to embed new production practices, but also to build new social relationships and to introduce new ideas into pre-existing rural organizations and social groups" (IPES-Food, 2018, p. 62) as part of their resistance to the mainstream. Some of the new ideas that early stakeholders were seeking to introduce involved organic production and marketing, as well as ideas about lifestyle balance and being in touch with nature (IPES-Food, 2018), which are consistent with the ideals of agroecology.

As an alternative to the status quo that at the time centered on industrial agriculture, stakeholders promoted and attempted to diffuse a new normative standard of eco-regionality. The emergence of this norm can be traced back to between 2002 and 2006 in the Limousin and Diois areas (EducEcoRegions Project, 2020). In essence, eco-regionality reflects values centered in the production and consumption of economically viable products that are ecologically and socially responsible, and are produced or processed as close as possible to their place of production (EducEcoRegions Project, 2020). In other words, the eco-regionality norm is the exact opposite of the foregrounded norms associated with the acceptance of industrial agriculture. The primary aim "was to restore the food sovereignty of the territories at a regional level by encouraging a logic of relocation and environmental sustainability with the advantages associated to organic production methods" (EducEcoRegions Project, 2020, p. 69). The principles and ideals of eco-regionality have been instrumental in adapting the sustainability norm to the local context of the Drôme. The Biovallée biodistrict, for example, has defined itself as a laboratory for innovation and experimentation, "working for sustainable development that aims to make the territory an eco-territory of reference while developing sustainable and livable activities" (EducEcoRegions Project, 2020, p. 70). The perpetuation of standards of behavior associated with the eco-regionality idea helps stakeholders to deeply embed the global sustainable development norm into local structures and institutions that consider "a whole territorial agri-food system" as opposed to a single supply or food chain, and shifts the focus to "the modes of coordination between various actors and institutions" (Bui & Lamine, 2015b, p. 1). The "paradigm shift" that eco-regionality requires is focused on the long-term sustainability and resilience of whole systems (EducEcoRegions Project, 2020, p. 67), and recognizes interconnections between environmental, economic, and ecological problems as opposed to viewing them in isolation.

The desired behaviors of local experimentation and invention are integrated into actions that center on "spread[ing] activities, services and innovative sectors adapted to [the] rural setting," (Biovallée, 2019, p. 3), and are folded into the overarching norm positioning the Biovallée as "a source of sustainable economic development" (Biovallée, 2019, p. 1). In this way, Biovallée stakeholders have called attention to the specific norms associated with the undesirable prior behaviors, while putting forth a new normative framework that helps to adapt the global sustainable development norm to the specific local context by calling for new standards of behavior via innovation and experimentation, which are positioned as the opposite of the foregrounded norms. The promoted eco-regionality norm is specific and requires specific reversals of behavioral expectations – a hallmark of the normative innovation strategy. Furthermore, Biovallée stakeholders have created indicators to measure and track actions underlying the eco-regionality norm, including the involvement of citizens in Biovallée actions, preservation and enhancement of the region's natural heritage via a small environmental footprint, innovative and transformative actions on a regional scale, and the structuring and development of a bioeconomy that produces a positive economic impact (Biovallée, 2019).

5.3.1 New Concepts and Transformed Perceptions

As Raymond et al. (2014, 202) explain, "Unless we contend that the 'stock' of extant norms is constant, it seems clear that the inventory of norms must change over time to include some new rules of behavior. New norms, as new standards of behavior, also often require the creation of new categories and concepts..." In other words, new concepts are an integral part of the normative innovation process. In the case of the Biovallée and other organic districts, the focus is on the novel concept of the ecoregion or biodistrict itself.

In this context, the biodistrict concept conceives of the territory as an ecosystem in itself. It likens the territorial ecosystem to the workings of a cell with its own metabolism (EducEcoRegions Project, 2020). This notion of "territorial metabolism" is the expression of "a very visionary political strategy since it integrates the intergenerational and multifunctional dimension of each action that is decided on its perimeter" (EducEcoRegions Project, 2020, p. 68). These concepts help to support the innovative norm of eco-regionality, on the basis of which stakeholders integrate the management of climate, energy, health, and food issues while working toward "the sustainable reconstruction of a region's immune system" (EducEcoRegions Project, 2020, p. 68). In practice, this means that all actions taken within the ecoregion must involve an assessment of environmental impact and must also consider long-term, integrated process streams from production to consumption.

The transition of the Drôme toward a biodistrict that disseminates the norm of eco-regionality is rooted in decades of resistance and active efforts by local actors working to implement new production practices, construct new social relationships, and introduce new ideas into longstanding rural organizations and social groups (IPES-Food, 2018). Sustained interaction between organic and conventional farmers, which took place in key venues for sharing knowledge, was instrumental in changing widely held perceptions about organic farming, leading to a new and expanded understanding. Importantly, these interactions "also encouraged a shift in the perception of organic farmers from 'backward', 'lazy' or 'crazy' individuals to forward-looking innovators" (IPES-Food, 2018, p. 63). As one source describes, "in many departments in France, the image of the wacky organic farmer persists, where organic farmers have fields full of weeds and diseases, no yields, where it doesn't work. In the Drôme, it's the complete opposite. For conventional farmers, organic is the most technically advanced, the best approach. ... Here, you frequently hear

conventional farmers say 'I am not good enough to engage in organic production'. ... But they still increasingly use at least some organic practices" (Bui, 2015,
p. 343). The transformed perception of local organic farmers as forward-looking
innovators highlights their instrumental role in normative innovation processes that
have served to translate the broader sustainable development norm to the local context. The local organic farmers of the Drôme can thus be viewed as the primary
norm entrepreneurs who took on the significant tasks of creating, reinforcing, and
spreading ideas about eco-regionality.

5.4   Conclusion

This chapter has analyzed the case of the Biovallée biodistrict, the only biodistrict
in France. It began by introducing the Biovallée and presenting a brief history of its
origin and development. It then focused on identifying the diverse actor constellations involved in the governance and management of the Biovallée. From here, the
focus of the chapter shifted toward environmental norms, first by distinguishing
prominent global environmental twilight norms that are reflected in discourse pertaining to the Biovallée, and secondly by evaluating the processes that local stakeholders, and particularly organic farmers, have used in adapting these norms to the
local context of the Biovallée. Overall, the analysis shed light on relevant norm-
based processes and concepts that bridge the global to the local setting.

Several findings have emerged from this analysis of the Biovallée biodistrict. To
begin, the norm of sustainable development is the predominant global norm in the
discursive field of the Biovallée, whereas the norm of intergenerational equity is
present but far less pervasive. No other global twilight norms are apparent in discourse related to this biodistrict. The sustainable development norm presents itself
particularly strongly in the context of discourse surrounding the aims and objectives
of the Biovallée, and in discourse focused on the Biovallée as a hallmark for regional
and global sustainability standards. In terms of the intergenerational equity norm, it
is mainly referenced in the context of resistance on the part of early and more recent
inhabitants of the region whose aim has been to protect and preserve nature, which
has markedly shaped the development of the Biovallée.

Relatedly, the chapter shows how stakeholders (particularly organic farmers)
have used normative innovation as a way of adapting the sustainable development
norm to the local context of the Biovallée. Again, the idea of resistance plays a key
role in this context. Stakeholders have foregrounded the standards of behavior that
support existing systems of industrialized agriculture, calling attention to the loss of
local food cultures, biodiversity, and local resources. Moreover, their resistance to
the mainstream has also involved the introduction of a novel normative framework
centered on the norm of eco-regionality, which organic farmers and other Biovallée
actors have perpetuated, and which has helped to translate the sustainable development norm to the specific context of the biodistrict.

New concepts and transformed perceptions are also part of the normative innovation process. This chapter thus places the biodsitrict concept into the broader theoretical context of normative innovation and norm adaptation, while highlighting conceptions of the regional territory as its own ecosystem. It also sheds light on the role of sustained interactions between organic and conventional farmers for the purposes of sharing knowledge in ushering in a positive transformation in local perceptions of organic production.

As normative innovation requires ongoing communication between stakeholders (Raymond, 2016), these early interactions between conventional farmers and the "innovative new inhabitants" (Bui & Lamine, 2015a, p. 7) that were settling in the area with environmental values that aligned with agroecology were likely critical in transmitting ideas about resistance and perpetuating the novel norm of eco-regionality in this territorial context. Moreover, opportunities for deliberation among diverse stakeholders, including citizens, government officials, farmers, and business owners, that stem from the participatory governance process of the Biovallée advance the deliberation process by providing opportunities to develop and refine new ways of thinking about agricultural practices and problems that have implications for territorial development. During the early development of the Biovallée, all of the area's inhabitants and stakeholders were invited to participate in a large forum in 2009 (Bui & Lamine, 2015a). Through deliberation that took place in a series of workshops and fora spanning 2 days, 200–300 people from across the entire Drôme valley provided their views on what the Biovallée should be like (Bui & Lamine, 2015a). Such strong attendance underscores "that there were strong expectations on the part of local actors and inhabitants to get involved in the project" (Bui & Lamine, 2015a, p. 13). The social movement literature details numerous examples of groups using collective discussion to model new social practices (Mansbridge & Morris, 2001; Weldon, 2002, 2006). Deliberation is thus critical to the development of novel norms, and it often first takes place in small groups – such as the fora described above – prior to efforts at spreading broader awareness (Raymond et al., 2014). Through discussion and deliberation such as the early Biovallée workshops provided, new sets of rules can come to govern behaviors and actions in ways consistent with the creation of a new normative framework. In these ways, processes of normative innovation have served as important mechanisms through which stakeholders have contextualized the broader sustainable development norm and adapted it to the specific local environment of the Drôme valley.

References

Assaël, K. (2017). *Creating a system of bio-districts in Italy within the national policies.* Retrieved from https://kipschool.org/usr_files/generic_pdf/BiodistrictsItalia2017-ENG.pdf
Basile, S. (2017). *The experience of Bio-districts in Italy.* Retrieved from https://www.fao.org/family-farming/detail/en/c/1027967/

Basile, S., & Cuoco, E. (n.d.). *Territorial bio-districts to boost organic production*. Retrieved from https://www.ideassonline.org/innovations/brochureView.php?id=91

Bio-distretto Cilento Association. (2019). *The Cilento bio-district*. Retrieved from https://www.ecoregion.info/wp-content/uploads/2019/09/CILENTO-brochure_EN.pdf

Biovallée. (n.d.-a). *La vallée de la Drôme*. Retrieved from https://biovallee.net/projet-biovallee/#vallee

Biovallée. (n.d.-b). *Un peu d'histoire*. Retrieved from https://biovallee.net/projet-biovallee/

Biovallée. (n.d.-c). *Un éco-territoire de référence*. Retrieved from https://biovallee.net/projet-biovallee/#ecoterritoire

Biovallée. (n.d.-d). *Une tradition d'actions locales*. Retrieved from https://biovallee.net/projet-biovallee/#tradition

Biovallée. (n.d.-e). *Les origines de Biovallée*. Retrieved from https://biovallee.net/projet-biovallee/#tradition

Biovallée. (2014). *Protocole grand projet Rhône Alpes Biovallée : Bilan au 9 juillet 2014*. Retrieved from http://drome-ecobiz.biz/upload/docs/application/pdf/2014-07/bilan_gpra_biovallee.pdf

Biovallée. (2019). *Présentation : Territoire d'innovation*. Retrieved from https://biovallee.net/wp-content/uploads/2019/07/Pr%C3%A9sentation-Biovall%C3%A9e-Tiga-11-Janvier.pdf

Bui, S. (2015). *Pour une approche territoriale des transitions écologiques – Analyse de la transition vers l'agroécologie dans la Biovallée (1970–2015)*. AgroParisTech.

Bui, S., & Lamine, C. (2015a). Full case study report: Biovallée – France. Retrieved from https://orgprints.org/id/eprint/29254/

Bui, S., & Lamine, C. (2015b). From niche to volume with integrity and trust: Case study fact sheet, Biovallée – France. Retrieved from https://orgprints.org/30703/1/Lamine_2015_Biovallee_Case%20study_FactSheet.pdf

EducEcoRegions Project. 2020. *O1-A1: Comparative analysis on organic districts (or eco-regions or bio-districts) in Europe*. Retrieved from https://www.ecoregion.info/wp-content/uploads/2021/11/O1-A1_Organic_Districts_in_Europe.pdf

IPES-Food. (2018). *Breaking away from industrial food and farming systems: Seven case studies of agroecological transition*. Retrieved from https://www.ipes-food.org/_img/upload/files/CS2_web.pdf

Mansbridge, J., & Morris, A. (2001). *Oppositional consciousness: The subjective roots of social protest*. University of Chicago Press.

Raymond, L. (2016). *Reclaiming the atmospheric commons: The regional greenhouse gas initiative and a new model of emissions trading*. MIT Press.

Raymond, L., Weldon, S. L., Kelly, D., Arriaga, X. B., & Clark, A. M. (2014). Making change: Norm-based strategies for institutional change to address intractable problems. *Political Research Quarterly, 67*(1), 197–211. https://doi.org/10.1177/1065912913510786

Weldon, S. L. (2002). *Protest, policy and the problem of violence against women: A cross-national comparison*. University of Pittsburgh Press.

Weldon, S. L. (2006). Inclusion, solidarity and social movements: The global movement on gender violence. *Perspectives on Politics, 4*(1), 55–74. https://doi.org/10.1017/S1537592706060063

Chapter 6
Conclusion: Looking Back and Ahead

This concluding chapter revisits the main research questions of this study and considers the key findings from across the three sub-national case studies presented in the previous chapters in light of those questions. In addition, it briefly reiterates the rationale for focusing on the sub-national level in France and identifies the value-added of this research. Finally, it extends the discussion beyond the country context of France by reflecting on how the theoretical concepts analyzed in this book could be used to examine how other types of global-level norms are being adapted and translated within other sub-national contexts, which can serve as the basis for future research in this area. Taken together, the areas of focus discussed in this chapter can be viewed as a framework for advancing scholarship on the adaptation of global-level norms in local contexts.

6.1 Research Questions, Key Findings, and Discussion

The primary focus of this study has been on the extent to which and how a specific set of global-level environmental norms – so called twilight norms – are being adapted to three local settings in France that each involve multi-stakeholder governance. The driving concern has been on how specific theoretical concepts and ideas can help us better understand how global environmental norms are (re)interpreted by local-level actors and translated to a particular local context. The primary assumption of this research is that local-level discourse in each of the three cases does not exist in isolation from the broader global-level environmental norms that have informed international environmental jurisprudence.

The first research question asked: Which of the global twilight norms are adapted to the local settings examined here? This analysis has shown that there is not a broad range of twilight norms that are evident at the local level. Rather, in each of the three cases, stakeholders concentrate on one or two key global twilight norms that they

M. Schnyder, *Global Norms in Local Contexts*, SpringerBriefs in Political Science, https://doi.org/10.1007/978-3-031-41108-3_6

adapt to the local setting. The sustainable development norm is the one twilight norm that is common across all of the cases examined here and as such is the most predominant norm in the analyses. One reason for this may be due to the ease of translating this particular norm to different local contexts relative to some of the other twilight norms, such as the common but differentiated responsibilities norm or the polluter pays norm. A second reason likely has to do with the sustainable development norm's explicit connection to the United Nations (UN) Sustainable Development Goals (SDGs). According to the SDG Index,[1] France currently ranks high at #7 among the 193 UN member states in terms of total progress in achieving all 17 SDGs, indicating that the sustainable development norm has likely been strong for quite some time in this country context. Other twilight norms that were less predominant but still present in the analyses include the intergenerational equity norm (which was present in two cases: the Cerbère-Banyuls Marine Nature Reserve, and the Biovallée biodistrict), and the equitable utilization of shared natural resources norm (which was present in the case of the Thau FLAG). The equitable utilization norm in particular fit well in the context of the Thau FLAG since the FLAG involved coordination among various regions, departments, and municipalities that shared in the governance and management of the Thau Lagoon, but one may not necessarily have expected this norm to translate equally well to other local contexts.

In focusing the analysis on twilight norms, one goal of this study has been to shed light on the continued significance of these global environmental norms, helping to address Beyerlin's (2008) observation regarding the confusion over the role that these twilight norms play in environmental governance and the effect that they have. The cases examined here have illustrated that at least several of these norms remain relevant in local-level environmental governance and management, and that local actors adapt and (re)frame these international environmental norms to make them relevant in local contexts. Although each twilight norm may not be equally prominent, several were strongly relevant in local discourse and policy frameworks. Overall, their continued relevance can be thought of as depending, at least in part, on how local stakeholders and activists continue to utilize international environmental treaties, standards, and objectives – such as the 17 SDGs – as a basis for integrating environmental norms on the local level.

Beyond assessing norm significance by identifying which norms are evident across the cases, this research also asked how local actors involved in environmental governance and management adapt and translate these global-level twilight norms to their local settings. One of the key findings here is that when we look at the local level, we see that local stakeholders have diverse ways of adapting the same global norm within the same country context. For example, although the sustainable development norm was evident across all three cases, the norm was adapted to each local context in quite different ways. For example, stakeholders associated with the Cerbère-Banyuls MNR and with the Thau FLAG both use grafting and norm

---

[1] The SDG Index is available here: https://dashboards.sdgindex.org/rankings

linkages, but in very different ways. Those active in the Thau FLAG also use normative reframing, as the Terre et Mer project and the Seafood Baskets project illustrated. In the case of the Biovallée biodistrict, stakeholders have relied more on normative innovation to discredit pre-existing norms that supported an undesirable status quo, and to perpetuate the norm of eco-regionality instead. This underscores that we can expect to observe many different ways of translating the same norm, even in a country context where congruence is high and there is a good fit between national and global environmental norms. This is because, for one, local contexts, values, economies, histories, and visions of the future can be quite different from one another, even within the same country. Secondly, different sub-national stakeholders and actor constellations can potentially interpret the same norm differently depending upon their interests and the context in which they operate, which opens the possibility to observe differences in processes of norm translation and adaptation (Eimer et al., 2016).

This study has also examined how global twilight norms are used in framing environmental issues across the three cases. It has highlighted how framing processes are specific to the local context, history, and identity. In the Cerbère-Banyuls MNR, for example, coexistence framing and the concept of patrimoine naturel are used to position the MNR as an important part of the natural heritage of France, and to articulate a desire to maintain this for future generations. Diagnostic and prognostic framing are also often used together in this context to construct a common understanding of how the MNR helps to safeguard biodiversity and the natural environment. In addition, the case of the Thau FLAG highlights the importance of issue frames and frame construction in positioning sustainable development as an integral part of the preservation of the maritime cultural heritage of the Thau region. Similar to those in the Cerbère-Banyuls MNR, Thau FLAG stakeholders also draw upon the concept of patrimoine in positioning sustainable development as a local matter with high stakes, involving preservation in both a cultural and economic sense. Finally, in the last case, we saw how organic farmers have used resistance framing in their efforts to foreground standards of behavior supporting industrialized agriculture, and to help introduce a novel normative framework that focuses instead on eco-regionality.

Taken together, stakeholders across all three cases reflect on and sometimes rethink environmental conservation and management in light of some of these global twilight norms, and in the context of local narratives and struggles. The cases have shown how environmental conservation, preservation, and sustainability actions can enable broad participation in planning and implementation, building locally focused "cultures of nature" (Chan et al., 2016; Light, 2006). Durable efforts at sustainability and conservation are still "often thought of as something imposed on local peoples by outsiders" (Chan et al., 2016, p. 1464). At the sub-national level in the three cases examined here, however, local stakeholders negotiate and translate global environmental norms collectively, which could potentially strengthen the legitimacy of conservation and preservation efforts "by engaging relationships with nature, with people through nature, and vice versa" (Chan et al., 2016, p. 1464), and potentially result in more effective and inventive sustainability efforts.

Overall, local stakeholders do negotiate global-level environmental norms via many different processes, underscoring the relevance of a diverse set of theoretical concepts and processes that are not often examined together in processes of norm adaptation and translation. Each case has demonstrated how norm translation involves complex processes involving local actors building congruence between the global norm and local values. Thus, this research does not conclude that it is reasonable to expect a static fit between global norms and local values (Checkel, 1999; Young, 1999; Cardenas, 2007), even in country contexts where congruence is high between global and national environmental norms. It instead supports the view that local actors, or "inside proponents" (Acharya, 2004), are heavily involved in congruence processes by engaging in dynamic actions to adapt global environmental norms to very specific and unique local environments in a way that strongly aligns with the identities and values that are relevant in that specific context.

6.2   A Focus on the Sub-National Level

The country context of France, overall, is characterized by a high level of congruence between global-level and national-level environmental norms. However, this research takes the position that even in such cases it can be worthwhile to examine variation at the sub-national level. I have argued that even where congruence is high, it is not perfect. Although most of the localization literature examines cases of low congruence between international and domestic norms, this research has shown that dynamics of norm adaptation and translation can also be expected to occur in cases of high congruence. This is because, in general, pure norm adoption involving no significant modifications is not likely to be observed (Eimer et al., 2016), which opens up possibilities of observing processes of norm adaptation and translation across a broader range of cases. As international norms embody a degree of ambiguity by design, domestic actors can translate them to the local setting according to the preferences and perspectives of the relevant stakeholders. With this in mind, this study contributes to the literature on localization and norm adaptation by examining theoretical concepts and processes that have typically been studied in isolation, including foregrounding, normative reframing, and normative innovation, as well as grafting and framing processes, under a common conceptual framework based on the notion of norm translation. This research has shown that norm translation to local contexts entails a range of dynamic processes. In addition, it has highlighted how at the sub-national level, different local stakeholders have varying interpretations of a given global-level norm, and although stakeholders may sometimes use similar dynamic processes to adapt those norms, the character and content associated with those dynamics differs by locale.

Consistent with the literature on norm diffusion (Wiener, 2009, 2016), this research has demonstrated that the subjective meaning attached to a particular norm will change to some extent as the norm is translated and applied in different local contexts involving different stakeholders, as norms can be expected to transform and evolve through processes of interaction. It has underscored that global environmental norms are mediated by context-specific local identities and structures, as well as by pre-existing local norms. Overall, this research suggests that in order to more fully understand how global norms are adapted to local contexts, future research should expand beyond cases of low global to national norm congruence and examine more cases where congruence is high. By examining such cases, we are more likely to observe the norm(s) in question across different local settings, which can facilitate comparisons of norm adaptation within the same broader institutional context.

6.3   Relevance for Other Contexts and for Future Research

The analytical framework used in this study to understand how global environmental norms are being adapted to local contexts in France can be useful for studying other cases, institutional contexts, and issue areas, providing opportunities for future research that could generate important findings and deepen our knowledge of processes of norm adaptation and translation. A few of those areas that would provide a promising foundation for future research are highlighted in this concluding section.

For one, future research could extend the analysis beyond environmental twilight norms to focus on other types of global norms, and attempt to explain variation in the extent or nature of their adaptation. For instance, research has shown how human rights norms have diffused at the international and domestic levels (Greenhill, 2010), but there has been less research that systematically examines how local actors adapt these norms in sub-national settings.[2] Even in country settings where congruence can be considered high between global and national human rights norms and ideals, such as in advanced democracies, it could be worthwhile to analyze sub-national variation in human rights norms adaptation. For example, why have human rights norms been adapted more strongly in certain local areas of the United States as opposed to others? What factors explain that variation? In addition to human rights norms, norms pertaining to women's rights and gender equality would be relevant to examine. Why, for example, within some democracies do we see women's rights flourishing in some local contexts and being restricted in others? Other aspects of environmentalism could also be examined in light of the global climate crisis and international norms that are emerging from international contexts such as the United Nations Framework Convention on Climate Change and the United Nations Climate Change Conferences. Beyond twilight norms, other relevant environmental norms might include climate justice norms, and the intersection between these norms and emerging equity norms.

---

[2] But see, for example, Och, 2018.

Keeping with a focus on global environmental twilight norms, future research could analyze one or two specific twilight norms in greater depth, examining their adaptation in specific local settings. For example, there has been some research on the polluter pays norm in the regional U.S. context, and how a group of states was successful in making private corporations compensate the public for their emissions by reframing the issue as a matter of public benefits and emphasizing the public ownership aspect of the atmospheric commons (Raymond, 2016). In-depth case studies could also examine the intergenerational equity norm, which tends to be strongly invoked by stakeholders at the local level with an emotional investment in a particular natural landscape or ecosystem, or by individuals who otherwise have a strong sense of place attachment to a specific natural environment (Gustafson, 2001).

Such studies could assess the ways in which particular types of framing (diagnostic framing, prognostic framing, and motivational framing) (Benford & Snow, 2000) or dynamic framing processes (normative reframing, normative innovation) (Raymond et al., 2014) have played a role in adapting certain twilight norms in different local settings. Since frame resonance, or the success of a particular frame, is influenced by these different types of framing, future research might look at how successful ways of framing based on a particular twilight norm identify the problem or issue and attribute blame or causality for it; identify solutions to a given problem or issue while identifying resources, targets, tactics, and strategies; and mobilize action around an issue.

Lastly, country contexts other than France would also extend opportunities to apply the theoretical concepts and processes examined in this book to global twilight norms. For example, the analysis could be extended to other democracies, potentially analyzing local-level norm adaptation in other settings where congruence is high between global and national norms. Local actors working in the context of advanced democracies should not be assumed to undertake norm adaptation in the same ways, as different institutional contexts may reflect different institutional cultures. Moreover, democracies in different world regions may vary in their political contexts in ways that are relevant to how local stakeholders translate global norms to their distinct local environments. Future studies might therefore examine if we can observe any variation in the specific translation processes and dynamics, and in how norm-based strategies are used in adapting certain types of global norms. Over time, these cases could be compared to cases where congruence is low to determine if systematic differences can be observed in how stakeholders adapt and translate global norms across these types of settings.

The theoretical processes and concepts examined here reflect a set of dynamic actions that stakeholders can use to adapt global norms. This research has attempted to bring together and examine dynamics that have typically been studied in isolation, including foregrounding, normative reframing, and normative innovation on the one hand, and framing and grafting on the other, under the common conceptual framework of norm translation. Although this research has focused on the topic of global environmental twilight norms, more research is needed to explore how local stakeholders dynamically adapt a range of global-level norms across different institutional contexts. Such research could help shed light on the conditions under which certain translation processes are used. Overall, in addition to the international and the national, local dynamics should be further examined in order to more fully grasp the significance of international norms. This includes international environmental norms, which are highly significant during this critical time when nation-states must grapple with the consequences of climate change. Sub-national dynamics thus represent an important step on the path toward the realization of environmental benefits, particularly in the absence of sustained national-level action.

References

Acharya, A. (2004). How ideas spread: Whose norms matter? Norm localization and institutional change in Asian regionalism. *International Organization, 58*(2), 239–275. https://doi.org/10.1017/S0020818304582024

Benford, R. D., & Snow, D. A. (2000). Framing processes and social movements: An overview and assessment. *Annual Review of Sociology, 26*(1), 611–639. https://doi.org/10.1146/annurev.soc.26.1.611

Beyerlin, U. (2008). Different types of norms in international environmental law: Policies, principles, and rules. In D. Bodansky, J. Brunnée, & E. Hey (Eds.), *The Oxford handbook of international environmental law* (pp. 425–448). Oxford University Press.

Cardenas, S. (2007). *Conflict and compliance: State responses to international human rights pressure.* University of Pennsylvania Press.

Chan, K., Balvanera, P., Benessaiah, K., Chapman, M., Diaz, S., Gómez-Baggethun, E., et al. (2016). Why protect nature? Rethinking values and the environment. *Proceedings of the National Academy of Sciences, 113*(6), 1462–1465. https://doi.org/10.1073/pnas.1525002113

Checkel, J. T. (1999). Norms, institutions, and national identity in contemporary Europe. *International Studies Quarterly, 43*(1), 83–114. https://doi.org/10.1111/0020-8833.00112

Eimer, T. R., Lütz, S., & Schüren, V. (2016). Varieties of localization: International norms and the commodification of knowledge in India and Brazil. *Review of International Political Economy, 23*(3), 450–479. https://doi.org/10.1080/09692290.2015.1133442

Greenhill, B. (2010). The company you keep: International socialization and the diffusion of human rights norms. *International Studies Quarterly, 54*(1), 127–145. https://doi.org/10.1111/j.1468-2478.2009.00580.x

Gustafson, P. (2001). Meanings of place: Everyday experience and theoretical conceptualizations. *Journal of Environmental Psychology, 21*(1), 5–16. https://doi.org/10.1006/jevp.2000.0185

Light, A. (2006). Ecological citizenship: The democratic promise of restoration. In R. Platt (Ed.), *The humane metropolis: People and nature in the 21st century city* (pp. 169–182). University of Massachusetts Press.

Och, M. (2018). The local diffusion of international human rights norms: Understanding the cities for CEDAW campaign. *International Feminist Journal of Politics, 20*(3), 425–443. https://doi.org/10.1080/14616742.2018.1447312

Raymond, L. (2016). *Reclaiming the atmospheric commons: The regional greenhouse gas initiative and a new model of emissions trading.* MIT Press.

Raymond, L., Weldon, S. L., Kelly, D., Arriaga, X. B., & Clark, A. M. (2014). Making change: Norm-based strategies for institutional change to address intractable problems. *Political Research Quarterly, 67*(1), 197–211. https://doi.org/10.1177/1065912913510786

Wiener, A. (2009). Enacting meaning-in-use: Qualitative research on norms and international relations. *Review of International Studies, 35*(1), 175–193. https://doi.org/10.1017/S0260210509008377

Wiener, A. (2016). Contested norms in inter-national encounters: The 'turbot war' as a prelude to fairer fisheries governance. *Politics and governance, 4*(3), 20–36. https://doi.org/10.17645/pag.v4i3.564

Young, O. R. (1999). Regime effectiveness: Taking stock. In O. R. Young (Ed.), *The effectiveness of international environmental regimes: Causal connections and behavioral mechanisms* (pp. 249–280). MIT Press.